Essential Principles for Autonomous Robotics

Synthesis Lectures on Artificial Intelligence and Machine Learning

Editors

Ronald Brachman, *Yahoo! Labs*
William W. Cohen, *Carnegie Mellon University*
Peter Stone, *University of Texas at Austin*

A Short Introduction to Preferences: Between Artificial Intelligence and Social Choice
Francesca Rossi, Kristen Brent Venable, Toby Walsh
July 2011

Human Computation
Edith Law, Luis von Ahn
June 2011

Trading Agents
Michael P. Wellman
June 2011

Visual Object Recognition
Kristen Grauman, Bastian Leibe
April 2011

Learning with Support Vector Machines
Colin Campbell, Yiming Ying
February 2011

Algorithms for Reinforcement Learning
Csaba Szepesvári
2010

Data Integration: The Relational Logic Approach
Michael Genesereth
2010

Markov Logic: An Interface Layer for Artificial Intelligence
Pedro Domingos, Daniel Lowd
2009

Introduction to Semi-Supervised Learning
XiaojinZhu, Andrew B.Goldberg
2009

Action Programming Languages
Michael Thielscher
2008

Representation Discovery using Harmonic Analysis
Sridhar Mahadevan
2008

Essentials of Game Theory: A Concise Multidisciplinary Introduction
Kevin Leyton-Brown, Yoav Shoham
2008

A Concise Introduction to Multiagent Systems and Distributed Artificial Intelligence
Nikos Vlassis
2007

Intelligent Autonomous Robotics: A Robot Soccer Case Study
Peter Stone
2007

Essential Principles for Autonomous Robotics
Henry Hexmoor

ISBN: 978-3-031-00435-3 print
ISBN: 978-3-031-01563-2 ebook

DOI 10.1007/978-3-031-01563-2

A Publication in the Springer series
SYNTHESIS LECTURES ON ARTIFICIAL INTELLIGENCE AND MACHINE LEARNING
Lecture #21
Series Editor: Ronald Brachman, Yahoo Labs; William W. Cohen, Carnegie Mellon University; Peter Stone, University of Texas at Austin

Series ISSN   1939-4608 Print   1939-4616 Electronic

Essential Principles for Autonomous Robotics

Henry Hexmoor
Southern Illinois University

SYNTHESIS LECTURES ON ARTIFICIAL INTELLIGENCE AND MACHINE LEARNING #21

ABSTRACT

From driving, flying, and swimming, to digging for unknown objects in space exploration, autonomous robots take on varied shapes and sizes. In part, autonomous robots are designed to perform tasks that are too dirty, dull, or dangerous for humans. With nontrivial autonomy and volition, they may soon claim their own place in human society. These robots will be our allies as we strive for understanding our natural and man-made environments and build positive synergies around us. Although we may never perfect replication of biological capabilities in robots, we must harness the inevitable emergence of robots that synchronizes with our own capacities to live, learn, and grow.

This book is a snapshot of motivations and methodologies for our collective attempts to transform our lives and enable us to cohabit with robots that work with and for us. It reviews and guides the reader to seminal and continual developments that are the foundations for successful paradigms. It attempts to demystify the abilities and limitations of robots. It is a progress report on the continuing work that will fuel future endeavors.

KEYWORDS

autonomy, anthropomorphism, mobile robotics, paradigms, navigation, multirobotics

Contents

Preface

Autonomous robotics has had a relationship with science fiction that is deeply rooted in our instinct to understand ourselves as primary actors in the world. Conceiving of robots that are autonomous has compelled us to build models and paradigms that are biologically inspired. The current state of the art is multifaceted. On the one hand, we are equipping our machines (e.g., vehicles) with robotic gadgets (e.g., navigation tools) to make decisions on our behalf. On the other hand, we are relinquishing certain well-understood operations to robotic automation (e.g., manufacturing). We also aspire for robots that will exist in harmony with us.

This book results from several years of teaching and learning about the fundamental principles of robotics. I owe much appreciation to numerous students who contributed in direct and indirect ways in the development of the core material. Among these students, I acknowledge and appreciate Lonnie "Kit" Wilkerson, Joe Viscomi, and Mejdl Safron. Mr. Anthony Kulis has coauthored Chapter 13. Ms. Arwen McNierney provided editorial guidance. This book provides a fresh, introductory outline of major developments and includes pointers for further exploration. It can be used as an accessible guide to existing literature. Although it complements them, it does not replace detailed material in classic textbooks, for example, Arkin (1998), Bekey (2005), Dudek and Jenkins (2000), and Choset et al. (2005). Whenever possible, the reader is sent to seminal sources for further details. This book demystifies the tremendous hype that surrounds robotics in popular culture. It provides the beginner with a well-balanced college-level introduction. This book includes a chapter on robotic arms (manipulators) used in the manufacturing industry. It takes a language- and platform-agnostic approach to avoid duplication of existing material.

Advanced undergraduate students who wish to understand the basics of robotics are the primary target audience. For managers and investors, I offer a detailed capability statement that debunks the illusions of science fiction. For the motion picture industry and science journalists, I offer a glimpse of robotics in the next 25 years.

With the proliferation of robotic applications in every area of our lives, including combat, the ethics of robotics has recently attracted a lot of attention (Lin et al., 2011). Discussions on this subject are evolving rapidly and include deliberations on the rights and privileges for robots. We believe that ethics and moral considerations for robotics have not matured and a recapitulation should be postponed for another decade. Therefore, this book omits an examination of these issues.

This book is divided into four main parts. Part 1: Preliminaries contains six introductory chapters on classical and modern robotics paradigms. This material could be considered as a first course in robotics. Part 2: Mobility includes four chapters on navigation and mapping techniques in practice today (Choset et al, 2005). Part 3: State of the Art, with four chapters, reviews selected research at the leading research institutions. These chapters supply adequate detail to replicate research results (Thrun et al., 2005). Probabilistic robotics deals with interpreting sensory data for localization and mapping and is well covered in (Thrun et al., 2005), thus is omitted from this book. Beyond probabilities, soft computing techniques, including the latest applications of fuzzy logic, are outlined (DeSilva, 1995). Part 4: On the Horizon is an overview of robots that are specialized for applications in remote environments, including space, oceans, and the military theaters as well as robotics applications for medical treatment. Throughout, we also provide forecasts on challenges, breakthroughs, and robotics in the next 25-30 years. Each chapter is self-contained and can be read independently of others.

Henry Hexmoor
2013
Carbondale, Illinois

I dedicate this book to my loving wife, Dr. Dona Reese.

PART I

Preliminaries

CHAPTER 1

Agency, Motion, and Anatomy

Allow me to briefly indulge in science fiction, as it has been instrumental in inspiring the evolving work in robotics. Beyond commonplace personal digital tools such as the PDA, there is a movement afoot at leading institutions for the development of tools that help people with memory and reasoning called *cognitive prosthesis*. There is no doubt that we will continue to augment our biological bodies with synthetic limbs and cognitive prostheses, such as Google glasses.

In the not too distant future, we will share our planet with silicone-based beings possessing biological prostheses. Combined with medical advances, life spans will vary widely and individuals will have choices concerning continued sentience. Perhaps that date is many years from now. For identification and rapid information processing, demographics will include attributes that reflect individual fitness levels and capabilities. These attributes will determine the best matches for performing tasks. Individual preferences and activity-based roles will capture the range of functionalities available. For instance, an individual might be a plumber who prefers not to work for biological origin beings (humans). Another individual might lack the emotional capacity for certain interactions and therefore might be averse to or unfit for counseling their peers.

The notion of agency as an actor is more ubiquitous, and further blurs the distinction between an embodied person and an entity that produces actions in the world. In the foreseeable future, achieving high levels of uniformity and ubiquity of rights and privileges between humans and robots remains in the realm of science fiction, conjuring up moral and ethical dilemmas (Lin et al., 2011). Many ethical considerations are of increasing interest. This interest is, in part, driven by the increasing role of robots in the military, where they are able to commit a wide range of actions that can be harmful—or even fatal—to humans. The spectrum of social and political attitudes and positions will reflect the independence and allegiance of different people. These attributes will partly determine the power stratification among people. For example, well-connected people occupying powerful positions in society will have relatively high social power compared to their peers. I end my brief foray into science fiction.

Fueled by corporate funding in the Far East (e.g., Honda), as well as research investments by government agencies (e.g., NASA), an active research trend in robotic research has been the replication of human internal and external functionalities, such as the robotic legs, arms, and hands that mimic realistic humanoid movement. There exists, however, no single research collection source on the diverse range of mechanisms and programming of humanoids. On one hand, this research significantly contributes to our understanding of human anatomy. On the other hand, the resulting robotic applications are paving the way for the intermingling of humans and robots. For example, there is a nascent industry in service robots that perform dirty, dull, and dangerous (D3) —as well as tedious—functions in private homes (e.g., vacuuming) as well as in public places (e.g., museum tour guides). A specific niche area is space: robots relieve humans (both astronauts and space mission crews) of mundane and tedious tasks. A byproduct of this robotic proliferation is the vanquishing of the prevailing popular social phobia about robotic takeover.

With continued funding and research mandates, all technical challenges for the creation of realistic humanoid robots are surmountable. However, this author believes that there are limitations to the usefulness of realistic humanoid robots. Suppose a robot is created that possesses 100% human form and function. I'll call this robo-Sue. On the one hand, robo-Sue would suffer from the inefficiencies and imperfections of human beings in general, since humans are not the best designed for many common functions—both physical functions, such as locomotion, and mental functions, such as arithmetic and remembering. On the other hand, even if robo-Sue is superior to its human peers, when is robo-Sue needed? As a species, we want our leaders and trend-setters to be imperfect, like the rest of us. Perhaps robo-Sue could be a tireless entertainer or good conversationalist.

Robots connote notions of agency and autonomy to human minds. This is evident from numerous motion pictures, such as *The Day the Earth Stood Still* (see Figure 1.1). Agency and autonomy are common characteristics of all living animals. The idea that there are independently determined courses of action is, at its core, the concept of autonomy. Autonomy, in turn, is an essential component of the concept of agency and volitional action (Hexmoor et al., 2003; Trappl and Payr, 2009). In contrast to behaviors that are reactive to external stimuli (e.g., reflex behavior to protect yourself from accidentally touching a hot surface), volitional actions are explicit, deliberate actions an agent generates, largely motivated by independent, internal deliberation (e.g., expressing a decision to rest or recharge after a period of work) (Hershberger, 1989). Without autonomy, an agent, such as a thermostat, is merely an ascription of action to a device that, at best, produces a weak sense of agency. Perhaps the property of autonomy and the quality of agency can vary on some sort of scale in relation to their surroundings. In sum, roboticists are interested in robots as agents possessing autonomy.

Figure 1.1: The robot Gort from the 1954 science fiction film *The Day the Earth Stood Still*

Let's leave science fiction visions in favor of a discussion of the vast knowledge that we can learn from animal and human anatomy and their enormous direct inspirations for robot builders. Life has been characterized by *heredity* and *metabolism*, with details beyond our outline scope (Ayala, 2010). Single-cell creatures were the only animal species on Earth until about 500 million years ago. Dividing and copying mechanisms are responsible for heredity among living creatures, and DNA contains the instructions for reproduction. Metabolism is required for organisms to synthesize DNA.

Although the precise origin of life on Earth is still unclear, biologists have suggested that Darwin's theory of evolution accounts for the development of species and their anatomical diversity. According to Paleontologist Neil Shubin, supported by numerous biologists, including John Long, animal anatomy that enables movement is traced back to simple, aquatic fish-like animals (Shubin, 2008; Long, 2012). In fact, the evolution of all living creatures originated with simple animals like ocean sponges and coral, that can still be found in places such as the coats of southern Florida and Australia. This stationary coral uses tentacles that contract in order to bring nearby small fish and plankton into its stomach. Once the prey is digested, the stomach reopens and expels waste products (Woolfson, 2008). In the instinctual search for food, sponges took an evolutionary leap to disconnect from their rocky base and used the ocean currents to move toward better hunting grounds. Numerous subsequent evolutionary leaps are responsible for the growth of limbs, leading to ongoing parallel diversification in animal shapes and sizes—legs, arms, wings, fins, hands, feet, and tails—for successful interaction with their environments (Shubin, 2008).

Leaving the discussion of early evolution, in much more recent history, humans have evolved from apes (Taylor, 2010). According to archeologists such as the famed Timothy Taylor, humans have retained the least brawn of all apes, are weak, and cannot thrive well in nature with a lack of food and shelter. However, humans possess the most brain power of all apes, which has allowed them to use tools and develop culture and language, resulting in the development of modern conveniences. Human life on earth has lasted less than 10 million years (Ayala, 2010).

Let's consider locomotion. In their search for food, sponges began to use the oceanic currents for transportation. Evolution led to the formation of limbs, hands, and fingers for use in movement and securing food. Many, if not most, current animal species have a sense of up and down, front and back, and left and right. This gives animals the ability to deliberately orient themselves in the direction of movement. Fossilized worms, known as amphiox, were the earliest animals that possessed heads providing a rudimentary sense of orientation (Gee, 2008). Sea anemones, which are a variety of jellyfish, have a distinct front and back, allowing them to orient themselves beyond amphiox (Fautin, 1991).

All animals move—cheetahs faster, snails more slowly. Muscle contractions are the basis of movement in many, but not all, animal species. Some animal groups do not have any muscles at all, as they branched off from the evolutionary path before muscle cells evolved. Yet these animal groups—for instance, the sea sponges—are not incapable of movement. Sponges are able to contract without muscles, although it is unknown which cells in sponges are actually contracting.

Humans have had an enduring fascination with anthropomorphism and automation, as seen in the development of service robots (Nocks, 2007). We believe that our fascination is organic, in the sense of a collective psychological attitude preferring humans and attributing human qualities to others. In popular culture, this preference accounts for the predominance of humanoid robots over other forms, such as industrial robots.

We seem to strive to replicate humanoid mechanical machines as objects of curiosity, feats of engineering, and fully fledged employees in factories. Publicly and privately, Japan and Korea are leading the world in large-scale robot-building efforts with a bias toward humanoids.

To the extent and scope we perceive outside loci of independent cognition, independent control, and compliance with human well-being, we ascribe the human qualities of autonomy, agency, and animation (AAA qualities) to things that provide useful services to us. Google's self-driving and mapping vehicles possess high levels of AAA qualities, whereas iPhone's Siri, with its voice recognition, possesses a more moderate level of AAA qualities. For instance, Siri does not work well with idioms, dialects, or common synonyms. GPS voice directions in automobiles provide even lower AAA.

There are a finite number of ways animals move in the world. Robotics has been inspired to replicate locomotion modalities in robots. In the following section we contrast common modes of locomotion and provide brief details for the adopted types of robotic locomotion.

1.1   PREDOMINANT MODES OF LOCOMOTION

Flow is one form of movement. Swimming and flying animals move their internal body parts to benefit from buoyancy and the energy of surrounding currents to relocate. Aerostats are controlled gas balloons using flow movement that are often used for surveillance. Figure 1.2 shows an aerostat that is commonly used in military surveillance.

Figure 1.2: A photograph of an aerostat in use for military surveillance (image courtesy of TCOM, L.P.)

Appearing as wiggling, worms and some bugs may also move their internal body parts to *crawl* (see Figure 1.3). Using the friction between their body and a high-friction surface, such as a tree, trunk snakes slither and slide as shown in Figure 1.4. Many animals, including humans, use their legs (i.e., an appendage that makes contact with the ground) to *walk*, seen in Figure 1.5. Walking is superior to other modes of locomotion in harsh environments where the surface is uneven and varied in texture. Legs are also used for running and jumping, which provide further dexterity for handling difficult terrain. Imitating leg functionalities is more difficult than other modes of locomotion in nature. A disadvantage with leg locomotion is the relatively high-energy consumption it incorporates, as illustrated in Figure 1.6.

Figure 1.3: A common earth worm crawling

Figure 1.4: A common sliding snake

Figure 1.5: Three people walking forward

A group of scientists, headed by associate professor Dr. Michael Nickel of Friedrich Schiller University Jena (Germany), is looking into movement without muscles. The scientists from the Institute of Systematic Zoology and Evolutionary Biologists are especially interested in the question of which evolutionary forerunners preceded muscle cells.

A major consideration when comparing modes of locomotion is that they differ on levels of energy consumption at various speeds. Figure 1.6 shows that flow at high speeds is the most efficient mode of locomotion. Rail transit is the most efficient transportation mechanism at medium to high speeds. At very slow speeds, crawling and sliding are the preferred options for energy efficiency.

Walking and leg use are best at moderate speeds. Although wheels are more efficient than legs by two orders of magnitude, they are inferior to flow systems. In robotics, wheeled locomotion requires simpler and fewer motors as actuation mechanisms. The parsimony of design has resulted in the preference for wheels over legs. Nevertheless, there are a number of legged robots.

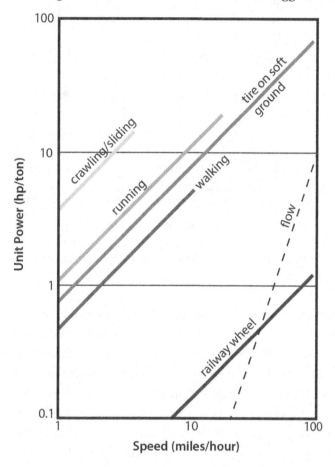

Figure 1.6: Energy consumption versus speed of locomotion (adapted from Siegwart and Nour-bakhsh (2011))

Beyond walking, Figure 1.7 illustrates the energy efficiency of bicycles over other modes of human transportation. In general, animal-powered locomotion is the far superior alternative for our health and has a small carbon footprint. Simplicity is also the key determinant of efficiency among motor vehicles. Larger and faster vehicles are bigger offenders in terms of fossil fuel consumption as well as pollution. Electric and hybrid vehicles are steps in the right direction, but their quantified impact is unknown.

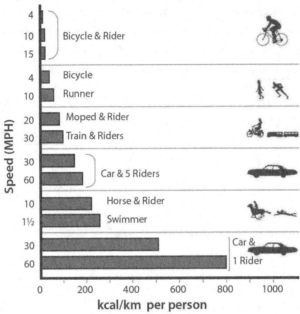

Figure 1.7: Energy cost of various forms of transportation (adapted from http://www.exploratorium.edu/cycling/humanpower1.html)

Often, modes of modern human transportation are useful in finite ranges, as shown in Figure 1.8. This figure illustrates conceptual comparisons among common modes of transportation. In the developed and developing worlds, pedestrian and automobile travel have long offered us a binary alternative. For very long distances, air travel is unrivaled. Bus and rail modes dominate public transportation in medium ranges. There are clear, marked inefficient markets in middle-range distances, such as between clusters of towns near large metropolitan cities. These population clusters are also the fastest growing magnets for dwellings (e.g. Tokyo and Los Angles). In large indoor spaces such as airports (e.g., Heathrow airport), there are new modes of point-to-point transportation. Heathrow has deployed transport pods that carry up to four passengers each, plus their luggage, on two-lane tracks at maximum speeds of 25 mph. Another promising development is car- and bicycle- sharing programs in certain metro locations (e.g., San Francisco), which is part of the *shared economy* movement in which individuals share their private houses, cars, and potentially other goods for small fees. This reduces idle times for expensive products, such as vehicles, with a positive byproduct for transportation. Dr. Shannon McDonald and her colleagues are also designing new modes of transportation such as the "travelator," a new generation of moving walkways at Heathrow airport.

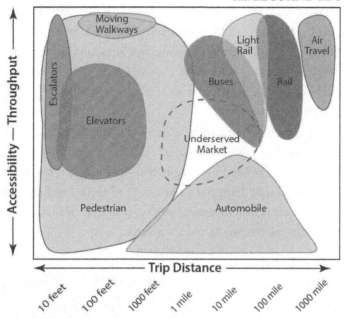

Figure 1.8: Conceptual comparison of modes of transportation in the Accessibility versus Throughput spectrum (adapted with permission from Young et al., (2007))

In the following section we will compare and contrast legged robots as a common mode of robotic locomotion.

1.2   LEGS AND LEGGED ROBOTS

Stability is the main concern for legged robots. Static stability is concerned with a robot's ability to stand without falling, when not in motion. Whereas one-legged robots are very unstable, two-legged robots are better at standing still without falling. Three legs are even more stable. Four-legged robots possess the most commonly known static stability. Dynamic stability is the robot's capacity to maintain an upright posture while in motion, and achieving dynamic stability requires continuous balancing.

The pattern of movement of animals, including humans, during locomotion over a solid surface is called *gait*. Let's assume legs are modeled as a single moving limb capable of only the actions of lifting and releasing (i.e., up and down movement). With basic walking, a gait can be simplified as a periodic series of lift and release actions for each leg. For a robot with k legs, $(2k-1)!$ is the number of possible actions to be coordinated among the legs (Siegwart and Nourbakhsh, 2011). With k = 2, this number, in which either of the legs takes action turns or both legs go up and down simultaneously, is 6. With 6 legs (i.e., k = 6), the number of possible actions jumps to 39,916,800. This point illustrates the complexity of accounting for the control of leg actions.

The first one-legged robot was built in 1983 by Marc Raibert at MIT, and it moved about by hopping on a single leg (Raibert, 1986). This robot was best suited for negotiating rough terrain. Obviously, it was not statically stable and could not stand upright unaided. However, since there was only a single leg, there was no need to coordinate leg actions with other legs. Hopping height, velocity of forward motion, and hip torque (i.e., up-rightedness) were its main control parameters (Raibert, 1986).

Two-legged robots are being pursued by electronic giants Sony and Honda, and each has built humanoid, bipedal, walking robots capable of climbing stairs. Two legs are more statically stable than one leg. They are lighter and more agile than multi-legged robots. However, dynamic stability is difficult, since successive footprint positions need to be carefully planned to maintain stability. Figure 1.9 shows images of humanoid, two-legged robots. Robonaut is a two-legged humanoid intended for assisting astronauts in space missions.

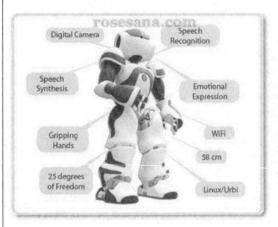

a.   b.

Figure 1.9: Images of humanoid robots. (a) An Aldebaran Nao humanoid robot (image courtesy of Alderaban.com) and (b) Astronaut and Robonaut shake hands on the ISS (image courtesy of NASA)

With four legs, as seen in the rugged, outdoor *alpha dog* image (Figure 1.10b), the robot is more stable than two-legged robots. In motion, it appears a bit clumsy from continual corrections. Genghis is a six-legged robot that is even more stable, see Figure 1.11. It has a tripod gait by considering the combined actions of three legs together. To determine what each leg should do, two legs on the opposing side of body are considered in order to maintain stability. Decentralized control is used to coordinate leg actions. Plague-infested fleas and beetles destroying crops are two examples that inspired the creation of robotic insects used in warfare (Lockwood, 2009). Research will continue to advance legged robots. There are also commercially available kits for building robots with a few or many legs, such as the centipede robot shown in Figure 1.12.

Figure 1.10: An alpha dog military robot (image courtesy of Boston dynamics)

Figure 1.11: "Genghis," a robot insect made at the Massachusetts Institute of Technology (MIT), USA. Credit: Peter Menzel/SCIENCE PHOTO LIBRARY. Used with permission

Figure 1.12: A centipede robot built from a kit from Mechano, Inc. (image courtesy of Mechano, Inc.)

Far more popular than legs, mobile robots employ wheels. Our last section in this chapter is devoted to a review of the variety of wheeled locomotion in more detail.

1.3   WHEELS AND WHEELED ROBOTS

A standard wheeled robot uses two or more wheels to produce locomotion. Figure 1.13 shows a robot with a wheel on each side, each separately and independently powered. This is an example of *differential drive*, in which adjusting relative power to the wheels' steering is achieved. The control program must determine rates and proportions of power to motors. An alternative is tricycle drive with three wheels (see Figure 1.14). As with the differential drive, two back wheels could be powered for propulsion and the third wheel used for steering. An option is to power the front wheel and use the back wheels as castor wheels for stability. Ackerman is a more complex type of drive mechanism, typically used in automobiles.

Figure 1.13: A Vex kit wheeled robot (image Courtesy of VEX Robotics, Inc.)

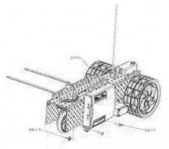

Figure 1.14: A Tricycle-drive robot (image Courtesy of VEX Robotics, Inc.)

Ackerman allows the inner turning wheel to turn at a larger angle than the outer turning wheel. Geometrically the front wheels of a car, if both are following a circular path around a common center point, must turn at slightly different angles. As shown in Figure 1.15, the inside wheel must angle in more than the outside wheel. If the turning wheels are not properly aligned they will fight each other, causing increased friction and wear. By increasing the inner wheel's turning angle, steering agility is greatly increased.

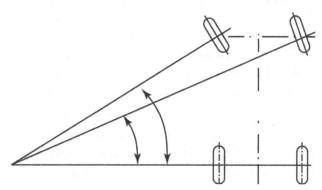

Figure 1.15: The Ackerman turning concept

Ackerman is used in motor vehicles. Ackerman steering provides a fairly accurate dead-reckoning solution while supporting traction and ground clearance. Generally, this is the method of choice for outdoor autonomous vehicles.

The *synchro* drive allows a robot to turn in place. In a synchronous-drive robot, each wheel is capable of being driven and steered. Typically, this is a system with two motors in three/four-wheeled drive configuration. One motor rotates all wheels to produce locomotion while the other motor turns all wheels to change direction and steering. Most often, there are three steered wheels arranged as vertices of an equilateral triangle, often surmounted by a cylindrical platform. All wheels turn and drive in unison. This leads to a *holonomic behavior*, which means it does not require performing complex actions for selecting a heading. Holonomic behavior is typically found in commercial robots since it is difficult to achieve in small-scale robots. A wide variety of other wheeled robots use tracks and other drive mechanisms to negotiate difficult terrain such as stairs, see Figure 1.16.

Figure 1.16: A stair-climbing robot built from a Vex robot kit (image Courtesy of VEX Robotics, Inc.)

1.4   CONCLUSION

The fascination with recreating biologically-inspired robots has led to broad diversification. There are people who pursue biological inspirations for robotics, including the recreation of locomotion (Lipson, 2008). Markus Fischer of Brighton University has created a bird-like flying robot with a six-foot wingspan. Hod Lipson of Cornell University has developed robots that adapt to locomotion without knowledge of their own anatomy. Whether we continue biological inspiration or not is an open question. Casting physical attributes aside, the neurobiology of animals has been an inspiration for the robotic architectures covered in Chapter 3. Seung tells us that we can uniquely identify ourselves by the configuration of our neural connections, which he calls *connectome* (Seung, 2012).

CHAPTER 2

Behaviors

In a popular paradigm that is termed *behavior-based robotics* (Arkin, 1998), observable, distinct interactions of a robot in an environment are likened to a finite set of distinct, self-contained, animal behaviors. This paradigm assumes the robot is an independent creature and that all of its interactions can be reduced to a mixture of distinct behaviors. These behaviors are either compared to select a winner (i.e., arbitrated), in which dominant behaviors take exclusive control of the robot, or blended, in which each behavior exerts a partial influence over the robot's interaction. We call this the *behavioral compositionality principle*, in light of its enduring popularity after its introduction by Rodney Brooks (Brooks, 1985). Ascribing behaviors to a robot, or even considering it an independent being capable of autonomous action, depends on the observer's perspective on which notions of agency are attributed to robots. We call this the *agency principle*. We will not elaborate on agency beyond the discussion in Chapter 1, as it falls outside of our scope.

There are forms of unintended behavior, called emergent behavior, that result from interactions among simpler components. Emergence is often found in swarms (Bonabeau, 1999). *Animat*, as used in artificial life literature as a synthetic, simulated animal, is another similar treatment of robots as animals (Franklin, 1995). What if there is a robotic being with components that each exhibit autonomous interaction? Extending the animal analogy will depend on contextual application.

By and large, animal behavior is innate and instinctive. It is the way animals react to internal or external stimuli through movements, postures, displays, eating, eliminating, mating, caregiving, hearing, smelling, hiding, threatening, killing, etc. Behavior varies among species. It varies among members of same species as well. It may determine fitness, i.e. the ability to survive and reproduce. It may also be determined by natural selection. There is ample evidence that behavior has a genetic basis (Plomin et al., 1994). Behavior-based robotics promotes modeling actions that appear to describe the biological actions of robot anatomy. Modeling animal behavior, akin to the study of ethology, has been a source of inspiration (Gould, 1982).

Selection is responsible for developing breeds with unique behaviors. Animals show behavior patterns unique to their breed, even when they are reared artificially. Some behaviors segregate in Mendelian fashion; i.e., crossbreeding among generations. Within a generation, behaviors may evolve to allow for improved fitness and progress along evolutionary path at the level of groups or individuals.

For an instinctive animal example, consider that all cows, everywhere, simultaneously face north (or south) while eating. Although its origins are not well understood, this is a distinct behavior. Magnetic alignment is the most common explanation for this benign and insignificant behavior (Begall et al., 2008). Similar behaviors are detailed in Balcombe (2010). For another, more mysterious animal behavior, consider that many animals, from cats to dogs to centipedes, will leave an area when they sense an impending earthquake.

Humans exhibit behaviors that are similar to animals. Some of these instinctive behaviors stem from self-preservation. Others stem from food seeking and reproduction. Yet other behaviors are cultural and learned, such as reciprocal greetings. Individuals are often not even aware of their own behaviors. Humans navigate by building cognitive maps, a behavior shared with rats. The cognitive map discovery is credited to Edward Tolman, and his research on rats' behavior when exploring mazes (Tolman & Honzik, 1930). He reported that rats often take the shortest path to the goal whenever possible and coined this behavior as a type of *spatial insight* (Tolman, Ritchie, & Kalish, 1946).

Another way we can understand behaviors is in pursuit of *Darwinian happiness* (Grinde, 2002). Grinde posits that behaviors are instruments in our evolutionary path. While sugar is delicious for human consumption, cows prefer something salty to eat (Grinde, 2002). Dietary requirements and preferences are evolutionary milestones. Fascination with unplanned animal behaviors, as well as computational sciences in the form of animal inspired, emergent behaviors, is well explored in *ethology* (Resnick, 1997). This emergence is also present in the makeup of the human mind (Minsky, 1998). Instinctive human behaviors can also be attributed to our endocrine system, which gives rise to homeostasis—the quality of living cells or organisms that allows them to maintain a stable internal state.

A large percentage of adult human behaviors are intentional in nature. In contrast, mundane human behaviors are not cognitive (i.e., deliberate) and therefore are more animalistic and unplanned. Whether we are satisfying biological and instinctive needs or accomplishing goals, most of our intentional behaviors are derived from reasoned out plans. Without our capacity to develop technology, our species would be extinct (Taylor, 2010).

Because of cultural training, followers of the Islamic religion face in the direction of Mecca at prayer times. A fraction of human behavior is not rational but is driven by subconscious moral principles. Emanuel Kant posits the moral law is grounded in the fact that it is governed by the human will and, as such, it is subconscious.

Valentino Braitenberg's seminal work (1984) describes fictitious vehicles that produce simple but interesting reactions to their environment. These reactions are rapid responses to sensed stimuli, as if they were reflexes generated by a nervous system of somewhat simple animals (e.g., a cat). Braitenberg developed thought experiments with the simple wiring of sensors to motors that produced behaviors an observer could recognize as meaningful behaviors. Figure 2.1 shows a simple vehicle called *Alive*. Alive has one sensor connected to one motor. The vehicle moves in response to the input data. The more intense the data, the faster the vehicle moves. If we poured a bucket full of Alive vehicles that are properly calibrated onto a hard surface (e.g., a comet surface during a space exploration of the comet) we would see the surface contour through the vehicles interaction with the surface.

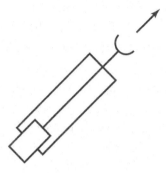

Figure 2.1: A Braitenberg "Alive" Vehicle

Braitenberg developed a simple model of robots with sensors and motors to show how non-trivial behaviors can arise from simple mechanisms. Figure 2.2 shows a differential-drive vehicle with two light sensors and two motors. Consider what happens when sensors are connected directly to the motor on the same side. This vehicle will likely spend more time in places where there is less light that excites its sensors, because it will speed up when exposed to higher light concentrations. If the light source is directly ahead, the vehicle may hit the source unless it is deflected from its course. This behavior is seen as *aggressive*. If the source is to one side, one of the sensors—the one nearer to the source—is more excited than the other, and its corresponding motor turns faster. As a consequence, the vehicle will turn away from the source. This behavior is seen as *cowardly*.

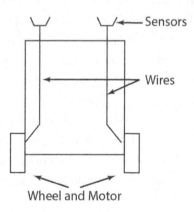

Figure 2.2: A simple Braitenberg vehicle with two sensors and two motors

Consider sensors that are connected through an inverter to the same side. Inverters give the opposite value that sensors do. If the source of the light is directly ahead, the vehicle will slow down as it approaches the source, unless it is deflected from its course. This behavior is seen as love. More complex logics concerning memory and state are available as quantum Braitenberg vehicles (Salumäe et al., 2012). Even more complex controllers can be developed with binary, multiple-valued, fuzzy, or mixed types. The behavior-based robotics approach stands in stark contrast to the development of the complex, monolithic approach to building a full-fledged humanoid cognitive system.

A parallel seminal work is that of MIT's Rodney Brooks, who led efforts in the construction of simple animal-behavior-driven robots. Brooks believes that roboticists should build simple behaviors, copying simple animals like ants, and then allow emergent behaviors to account for more complex behaviors, such as group work and collective functions in ant colonies. He argued that the distance between sensing and acting is small in behaviors. This is likened to the rungs of a ladder. Each rung is a behavior, with lower ones as faster behaviors and higher ones slower.

The remainder of this chapter will outline common behavior-based robotics techniques and methodologies in practice, as well as those considered in the classroom.

2.1 COMMON BEHAVIOR TYPES

Reflexes are fast and involuntary movements. Human infants possess rooting behavior that helps them locate nourishment. This reflex can be initiated by stroking the corner of a baby's mouth. She will turn toward the source in an attempt to find food. The rooting reflex subsides at around four months, as the baby becomes better at reaching and her movements become more voluntary. The Babinski reflex can be seen if you stroke the sole of a baby's foot firmly. Her big toe will bend backwards toward the body, and her other toes will fan out. This reflex will last in babies until they are about 2 years old. Other infant reflex behaviors are sucking, swallowing, the Moro "startle", grasping, and stepping. These subconscious behaviors remain with us throughout our lives.

Conscious behaviors, on the other hand, are deliberate and complex. The *relaxation response* is a largely conscious behavioral response to activities that produces a feeling of relaxation and lowered anxiety and stress. It can be invoked by therapy training methods such as meditation and massage.

Reflexes must either occur spontaneously as a part of the robot's routine activities or in response to a stimulus. For a robot, *backing up* from unintentionally contacted obstacles can be a reflex. *Veering away* can be another reflexive behavior if a moving obstacle is perceived to cross the path of the robot. A set of reflexive behaviors can be designed that completely account for a robot's obstacle avoidance patterns.

If a level of stimulus is needed to initiate the display of a behavior, it is said to be a *threshold-based behavior*. Some behaviors have a very low threshold; e.g, fighting dogs will attack with little provocation. A threshold-based behavior for a robot may be to monitor its battery charge and, if the charge falls below a certain level, trigger moving to the recharge station.

Latency is a parametric part of a behavior that produces a slower reaction to stimuli. The duration between reaching the threshold and triggering the response for a reflex is the latency, and it is parametrically adjusted.

Taxis are the instinctive tendency of an animal to approach or to avoid a particular stimulus, such as light and wetness, with successive corrections. Taxis differ from *tropism*, which is the turning response, often ingress toward or away from a stimulus. With taxis, the organism has motility and demonstrates guided movement toward or away from the stimulus source. *Klinotaxis* is a type of taxic behavior that performs successive comparisons for incremental adjustments to orientation. In other words, the use of continual feedback is required. Figure 2.3 shows an example.

Figure 2.3: An example of a taxic behavior to turn toward the light

Tropotaxis is a taxic behavior in which sensory inputs are compared in order to arrive at a decision. For example, consider Figure 2.4 where, in order to travel between two objects, perceptions of both objects must be compared in order to find the middle.

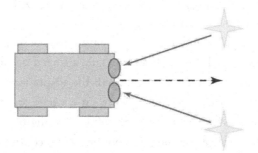

Figure 2.4: An illustration of a tropotaxic behavior to align with two light sources

Figure 2.5 illustrates *teleotaxis*, where the idea is to choose among sensory inputs that will produce alternate (possibly opposing) reactions. When presented with a number of target objects, teleotaxis determines the choice of which target to approach. One example is when facing a corner, which obstacle (i.e., wall) to turn away from, as is shown in Figure 2.5.

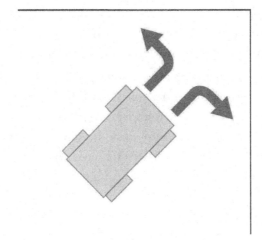

Figure 2.5: An illustration of a teleotaxic behavior of a robot confronting walls in a room corner

A *condition-based* taxic behavior maintains a condition during a response. For example, Figure 2.6 shows a station-keeping behavior in which a certain distance is maintained (i.e., a condition of constant distance between vehicles) when following a moving object.

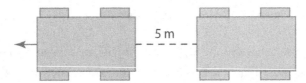

Figure 2.6: An illustration of condition-based taxic behavior, using an example of station-keeping behavior

Behaviors are often malleable and dynamic. Acquiring new behaviors can occur in the form of adaptive behavior, in which a behavior is fundamentally changed or a behavior's strength in response to changing environmental conditions is changed. The adaptation might be guided by innate instincts and the subconscious or conscious experience. In contrast to adaptation, learning in the computational setting is often accomplished through soft computing techniques including artificial neural networks, genetic algorithms, evolutionary computing, fuzzy control, and a hybrid mechanism.

Learned behaviors might be novel behaviors that are acquired through the recognition of (i.e., becoming aware of) a repeated pattern of behavior. Alternatively, learned behavior can be deliberate due to a myriad of machine-learning algorithms covered elsewhere (Kaelbling, 1993). Learning new behaviors is good for handling novel situations that might confuse older behaviors.

Homeostasis is an organism's innate system of control mechanisms that provide feedback essential for survival and comfort. It regulates internal functions such as temperature, humidity, etc. A related notion is *autopoesis*, which is homeostasis of identity.

Swarming behavior is a type of taxic behavior. Swarming behavior is the emergent collective behavior of decentralized, self-organized systems, natural or artificial. Biological swarms in nature are often not controlled. Control is possible if we can monitor and adjust the balance between inter-individual interactions and the simultaneous interactions of the swarm members with their environment (McLaughlan and Hexmoor, 2009; Gazi and Passino, 2003).

There are many common words used to describe swarms, including flocks, herds, and schools (Figure 2.7). Figure 2.8 shows a robot swarm in the European Symbrion project for designing and developing paradigms for pervasive robotic systems based on bio-inspired adaptation strategies that form robots into symbiotic organisms. These organisms are examples of complex robotic systems with distinct behaviors in which each robot component also exhibits distinct sets of behaviors that are separate from the organism.

a.   b.

Figure 2.7: Two examples of swarms in nature: (a) a flock of auklets, and (b) a herd of caribou

Figure 2.8: A robot swarm from the Symbrion project. From Kernbach et al. (2008)

2.2   CONCLUSIONS

Behavior-based robotics has inspired numerous activities from hobby robots to robust academic research. Its intuitive appeal is deceptively simple and allows for the development of a rich plethora of programs. An issue that arises is the need for organization among competing behaviors. One way to deal with this is to create a priority scheme (i.e., behavior arbitration strategy), as in Rod Brooks' subsumption architecture. Figure 2.9 shows a style of hierarchies over a set of behaviors. Level 0 behaviors can be overtaken (i.e., subsumed) by level 1 behaviors. Higher-level behaviors subsume lower-level ones. The lowest-level behavior is the robot's default behavior.

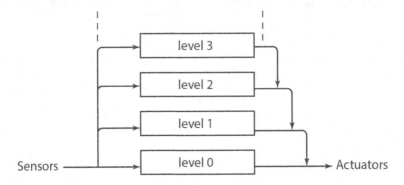

Figure 2.9: Subsumption among a set of behaviors imposes a priority among them (adapted from Brooks, 1985)

CHAPTER 3

Architectures

The experience of organizing codified behaviors and other frequently repeated, programmed procedures culminated in solidifying a set of principles that is generally termed *robot architecture*. It records robotcists' common concerns for arranging software in a coherent and endurable fashion, additionally employing software engineering styles (Hexmoor, et al, 1997; Hexmoor and Desiano, 1999).

Unlike other programs that are finite in duration, programs for robotic platforms are intended to repeat and run indefinitely, mirroring a living agent. One of the earliest conceptions of cyclic functions in a living creature is the *Sense-Think-Act* cycle, shown in Figure 3.1, that is a traditional artificial intelligence paradigm. It borrows from human cognition research. Inspired by control theory, the Sense-Think-Act cycle has, as the core goal, to continuously attempt to minimize the error between the actual state of the world and the desired state of the agent.

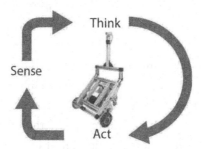

Figure 3.1: Sense-Think-Act cycle

As shown in Figure 3.1 and 3.2, Sense-Think-Act is an infinite loop. Figure 3.2 is a depiction of a robot (i.e., an agent) that continually senses and interacts with an environment. The robot *senses* to gather environmentally pertinent data. The robot *thinks* to make sense of the environment and determines what it can and should do, based on a comparison with its internal beliefs, desires, and intentions (Wooldridge, 2009). After deliberation, the robot *acts* to manifest changes in the environment and observes the effects as the cycle continues. The robot continually senses its environment, considers the relevance of what it finds to its goals, formulates a course of action through thinking, which in turn leads to it choosing and performing an action.

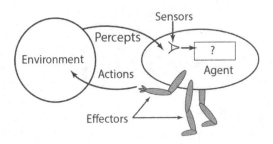

Figure 3.2: An illustration of a robot's interaction with the environment

This cycle establishes a steady loop that propels the robot toward its goals. In this context, sensing encompasses much more than gathering data. It includes both perception and the triggering of conceptual notions. For example, sensing obstacles from the front and the sides can trigger the idea of being trapped in a box canyon feature. Alternatively, sensing might trigger proprioceptive cues. For example, if the robot needs to travel a certain distance, the *proprioceptive* cue might be that it has reached the desired travel distance.

Finally, sensing may include belief revision. New sensory information can require changes in previously held beliefs about the state of the world. The internal representation of the external world is the robot's *world model*, which needs to be continually revised via sensing and updating data, beliefs, and information about conditions in the world. The world model commonly includes maps and locations on maps (i.e., localization). Classic mobile robot architectures prior to the 1980s are considered to be *pipeline architectures* (also known as *horizontal architectures*), in which outputs of sensing modules was fed to a model module to update the internal representation, then to a planning module to update or create a plan of actions, next to an execution module to command the sensors and motors, and finally to a control module to ascertain plans were completed.

The Sense-Think-Act cycle is a fundamental characteristic of conventional cognitive science. It holds an assumption that the embodied approach considers to be fatally flawed. The second main component of Sense-Think-Act is responsible for considering a robot's options, desires, intentions, and actions. Desired options are revised to be commensurate with sensory inputs. Robots deliberate about what intention should be achieved next. This may require revising and updating plans. The process used might entail means-ends reasoning to craft a plan to accomplish the intention. The act component can overlap with thinking, including revising intentions and selecting a single intention to manifest. Acting is also responsible for executing the plan. If the robot is engaged in lower-priority activities, the latest act will suppress them and will start control of the actuators. Before the robot can return to sensing, there needs to be a sufficient amount of time to allow the motors to complete their function.

Time is also needed for interaction with the world to produce intended changes. We'll call this the *pause* segment of the loop. During the pause period, the robot may need to communicate with the rest of its system. It may also generate and deliver feedback to users and other system components.

Figure 3.3 summarizes the sense-think-act function in a five-step infinite loop. This is reference architecture, which means that it is a high-level superset blueprint architecture. For a specific robotic system, the degree of detail at each step depends on the sophistication of a robot for that component. For instance, thinking for a simple, differential-drive robot that is designed to wander (wanderer) will only be a concern when selecting an orientation (i.e., a heading). Such a robot will not be building or revising an elaborate plan. On the other hand, for a robot that is unloading a cargo ship (unloader), thinking must include building and maintaining a detailed plan for the stages, phases, and sequences of actions that need to take place. While pause for a wanderer is simply inaction for a short duration, pause for an unloader will include communication and coordination with other unloader robots. For complex unloaders, coordination during pause might require ontological mediation to make sure they are considering work with similar (or even the same) cargo elements.

```
1. While true {
2. Sense the world—(a) sensors, (b) communication, (c) supervisor input
    1. Form perceptions—(a) concept triggering, (b) propioception
    2. Update beliefs (belief revision)
    3. Update internal world model—(a) map, (b) localization, (c) relationships and attributes
3. Think about options, desires, intentions, and actions
    1. Revise desirable options and select one
    2. Deliberate about what intention to achieve next
    3. Revise and update plan
    4. Use means-ends reasoning to get a plan for the intention
4. Act
    1. Revise intentions and select an intention to manifest
    2. Execute the plan
    3. Suppress less important behaviors
    4. Start control of actuators
5. Pause
    1. Until the world changes
    2. Communicate
    3. Generate and deliver user feedback
}
```

Figure 3.3: Top-level robot control loop reference architecture based on the Sense-Think-Act cycle

In 1985, MIT's Rodney Brooks revolutionized the horizontal paradigm by introducing the *vertical model* for organizing behaviors in what he termed *subsumption architecture* (Brooks, 1985, 1986). Figure 3.4 shows a set of behaviors. The bottom behavior (level 0) is wander, which is the robot's default behavior. Avoid obstacles is the second behavior (level 1), which will subsume (i.e., trump) the wandering behavior. If it is triggered, it takes exclusive control of motors to avoid obstacles. Under certain conditions, exploring (behavior 3, level 2) will subsume obstacle avoidance. When possible, map building (behavior 4, level 3) will subsume exploration behavior. These four behaviors are in competition to control sensors and motors.

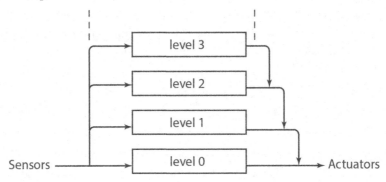

Figure 3.4. Subsumption Architecture (adapted from Brooks, 1985)

In the 1990s, there was a realization that some conceive and consider robot-world interactions at a conceptual, cognitive level (i.e., the high level) (Lammens et al., 1993). Others were considering robot-world interactions at a physical, engineering level (i.e., the low level). In order to accommodate both camps, hybrid architectures were born (Hexmoor and Shapiro, 1997). Later, three-tier architectures (3T) became dominant (Bonasso, et., al., 1997 and 1998; Connell, 1989).

3T architecture separates the general robot intelligence problem into three interacting layers or tiers. A skill manager coordinates a dynamically reprogrammable set of reactive skills. Sequencers, as the middle layer in 3T, activate and deactivate sets of skills to create networks that change the state of the world and accomplish specific tasks

The top layer of 3T is the Planner. It fulfills the duties of the mission planner and cartographer by setting goals and strategic plans. The deliberative planner layer reasons in depth about goals resources and timing constraints. Perhaps the most comprehensive cognitive architecture is the hierarchical, real-time control system (RCS) reference model architecture that is under development at the National Institute of Standards (Albus and Mystel, 1996). RCS is the reference model architecture that has been used over the past 25 years for a number of intelligent systems, including the NIST's Automated Manufacturing Research Facility (AMRF), the NASA telerobotic servicer, and the DARPA Multiple Autonomous Undersea Vehicle project. Figure 3.9 shows a number of cooperating robots controlled with RCS.

The goals set by the Planner are passed to the middle layer, the Sequencer. The Sequencer uses a reactive planning technique called RAPs to select a set of primitive behaviors from a library and develop a task network specifying the sequence of execution for the behaviors for the particular subgoal. The sequencer is responsible for the performance monitoring functions of a generic hybrid architecture. To complete the plan, a set of appropriate behaviors (skills) is instantiated. These behaviors form the bottom layer, the controller or skill manager.

An alternative three layer architecture shown in Figure 3.5 uses learning to migrate and codify routine activities into the high-level layer (Hexmoor and Shapiro, 1997).

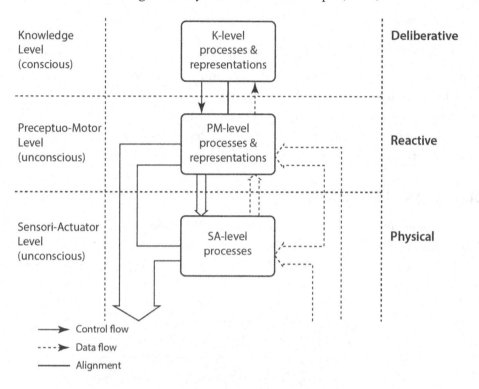

Figure 3.5: 3T architecture (adapted from Hexmoor and Shapiro 1997)

GLAIR (Grounded Layered Architecture with Integrated Reasoning) is a multilayered cognitive architecture for robotic agents operating in real, virtual, or simulated environments. The highest layer of GLAIR architecture (shown in Figure 3.5) is the knowledge layer (KL) that contains the beliefs of the agent and performs conscious reasoning, planning, and action selection. The lowest layer of the GLAIR architecture, the Sensori-Actuator Layer (SAL), contains the controllers of the sensors and effectors of the hardware or software robot. The layer between the KL and the SAL is the Perceptuo-Motor Layer (PML), which grounds KL symbols in perceptual structures and subconscious actions, contains various registers for providing the agent's sense of grounding in the environment, and handles translation and communication between the KL and the SAL. GLAIR was inspired by the tenets of human psychology and biology.

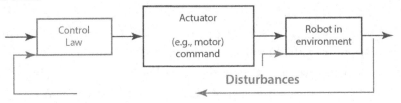

Figure 3.6: A feedback loop

Figure 3.6 shows a feedback loop, often used at the low level for robot control. The leftmost arrow shows sensory input entering the loop. Control law dictates the control regime, guiding the signals to be sent to the actuators. The most popular control laws are known as PID. P is used when the controller output is proportional to an error or change in measurement. I is used when the controller output is proportional to the amount of time the error is present. Finally, D (i.e., differential) is used when the controller output is proportional to the rate of change of measurement with time.

3.1   CONCLUSIONS

Robot architectures were popular topics in 1990s and ushered in many state-of-the-art, complex robotic systems. NASA mission robots included extravehicular, helper-retriever robots at NASA Houston, such as JPL's NASA MARS rover (shown in Figure 3.7) and NASA Moffet field (shown in Figure 3.8).

Figure 3.7: An artist rendition of the MARS Rover Robot. Photo Credit: NASA

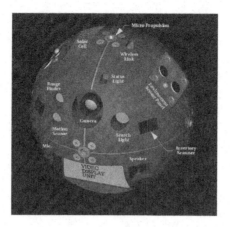

Figure 3.8: NASA's flying personal satellite assistant (PSA) robot designed to be used for inventory tasks on the international space station. Photo Credit: NASA

Figure 3.9: Autonomous underwater robots using RCS Architecture

CHAPTER 4

Affect

Affect is a phenomenon that manifests itself in the form of feelings (i.e., emotions). To be alive is to have feelings. There are countless named feelings among human languages for expressing affect. Four of the most universal feelings in English are happy, sad, fear, and awe. Skinner (1997) argued that all animal behaviors and all processes for behavior selection (i.e., choosing a behavior), including our own, are controlled by feelings. Degrees of preference or aversion for objects and behaviors can be used to measure our experiences of feelings. Emotions are largely non-rational. Although possessing emotions is a litmus test for sentience, spanning all animals, they are inadequate for characterizing consciousness. In addition to affect, the capacities for language and self-awareness are also required for consciousness (Chalmers, 1996).

While contemporary robots lack consciousness, we are moving in the direction of empowering them with the capacity to become conscious. In the future, we will witness robots who understand the differences between themselves and humans. A step on the path to consciousness is to model intentions and reason using intentional stance (Dennett, 1989). Robots need to become aware of the intentions of their human counterparts in order to reason about their underlying mental states of belief, desire, and goals, which will produce greater coherence and collaboration between humans and robots. Robots need the capacity to generate intentional behaviors and to work in environments where there is *mixed intentional interaction*, which is a concept that dates back to early artificial intelligence research (Carbonell, 1970), with roots in computational language (Haller, et al, 1999) as well as multiagent research (Horvits, 1999).

Examples of mixed initiation are abundant in routine interactions among collaborative teams where there are frequent interactions among individuals, such as in surgical procedures and mechanical repairs. In order to coordinate, individuals engage in rapid exchanges of requests (initiating and expressing intentions) and responses (honoring intentions generated by others). Future robots will engage in rapid interactions with us that are indistinguishable between humans and robots. This high level of flexibility requires nontrivial robot consciousness. For brevity and scope, we will limit the remainder of this chapter to an outline of affect as a subset ingredient for consciousness.

Affective Computing (AC) was coined by Rosalind Picard of MIT (Picard, 2007). Broadly speaking, AC straddles two thematic areas. In the first area, AC is concerned with computer programs that account for human-user emotions. In the second area, AC is used to produce synthetic feelings and enable robots to exhibit emotions during interaction with, or for, the entertainment of their human counterparts.

AC is motivated in a variety of ways. Humans often use emotions for decision making when there is an absence of rational and conscious motivation. Facts and principled decision making are often very slow and encumbered, whereas emotions are associated with learned reactions and lead to fast and fluid choices of behavior. For example, fear is the human emotional response to danger (e.g., fire), which leads to self-preservation actions (e.g., fleeing the fire zone).

To be of service and capable of social relationships, robots that interact with humans must possess the capability to detect and use emotions with their human counterparts. Cognitive models, such as the ones for learning, can be slow, brittle, and awkward. Emotions can be used for facilitating learning reactive behavioral responses (Power, 1992). The perception of loading and unloading cues guide and expedite other perceptions, such as visual perception. For instance, musicians use a multitude of complex, emotional cues. Mobile robots are used as therapeutic tools for children with autism who lack the ability to understand and demonstrate appropriate emotional expressions (Dautenhahn and Werry, 2004). In order to arbitrate among sets of cognitive functions, emotions are useful tools (Lorini et al. 2008). Emotions can help break ties among competing cognitive processes. AC explores the ability to send and receive emotional cues. In general, AC can enhance the quality of interactions between humans and computers.

With robots, emotions can be used for feedback and encouragement in human-robot interactions. Robots benefit from simple emotions for coordination when working with children or people. There is a nascent but rapid industry in service robots that successfully employs emotions. Figure 4.1 shows a service robot that can provide support for various services in offices, commercial facilities, and other public areas in which people work or spend leisure time.

Social informatics is another emerging discipline for the exploration of communication tools in various social and cultural contexts. Emotional intelligence is the quality of perceiving, assessing, and influencing ourselves and others. Naturally, emotional content will vary from person to person.

Figure 4.1: Fujitsu's service robot, called enon (image courtesy of Fujitsu.com)

AC is used in many robotic applications and is useful for augmenting collaboration between humans and computers. A primary robotics application is to produce anthropomorphism in human interfaces—giving a robot human-like emotions and behavior. Emotions are useful in performance monitoring and can enhance coherence in communication between humans and robots. They can be used to generate feedback or monitor the progression of emotions in others. There are several neurological models that explain the role of the Amygdala (shown in Figure 4.2) in the human brain.

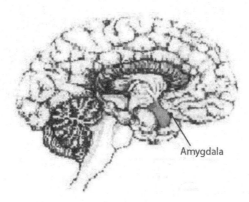

Figure 4.2: A conceptual image depicting the place of the Amygdala in the human brain

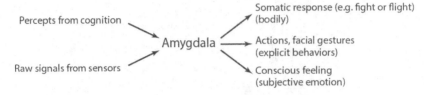

Figure 4.3: Amygdala Inputs and outputs (adapted from Wehrle and Scherer, 2001)

A neurological model of the Amygdala is shown in Figure 4.3 (Wehrle and Scherer, 2001). *Percepts* are mental concepts triggered by sensory data. Percepts as products of perception from cognition constitute a set of inputs to the Amygdala. Raw sensory signals are another set of inputs to the Amygdala. A type of output is *somatic responses*, which covers bodily responses such as sweating, rapid breathing rhythm, and increased heartbeat rate. Somatic responses are largely limited to the individual and not intended for communication with others. Another type of output is explicit displays of behaviors such as facial gestures. Although these also remain within the individual, they can be implicit forms of communication. Figure 4.4 shows an example of facial gestures in an automobile driving aid device clearly expressing the emotional status of the device, showing happy and sad with eyes drawn up and drooped down respectively. Yet another type of output is conscious feelings, which are predominantly subjective emotions that are intended and often explicitly expressed.

Figure 4.4: MIT's GPS enabled Driving Intelligent Agent called Aida exhibits emotions (adapted from Di Lorenzo et al. (2009))—Audi.com

There are many different types of emotions (Ortony et al., 1990). One class of emotions is *goal-based emotions* (GBE). This class can be divided into positive or negative groups. Experiencing joy for positive outcomes, having hope for positive outcomes, and relief from aversion of negative circumstances are all positive GBE. Negative goal-directed emotions for GBE include fear and distress caused by negative outcomes, and disappointment for not reaching a favorable outcome. A second class is *standards-based emotions* (SBE) that is largely normative or culturally characterized. Positive SBE is found in pride and gratitude toward oneself, as well as admiration and gratitude toward others. Negative SBE is shame and remorse toward self and reproach and anger directed toward others. The third class of emotions is *taste-based emotions* (TBE) that are endogenous and innate to individuals. Love and attraction toward things are positive TBE, whereas hate and repulsion from things are negative TBE.

By and large, emotions are subjective and not easily quantifiable. They are not crisp in nature and are often non-scalar, i.e., defying accurate measurement. Emotions occupy a spectrum. Their *valence* can range from positive to negative. Independently, an emotion will have an *intensity*, which is the value denoting the emotion's level of arousal,i.e., amplitude. Intensity can range from boredom to surprise.

Robots equipped with emotions are rapidly emerging in academia and industry. Figure 4.5 shows one of the earliest robots, named Kismet, interacting with Cynthia Brazeal. Kismet has a repertoire of responses driven by emotive and behavioral systems.

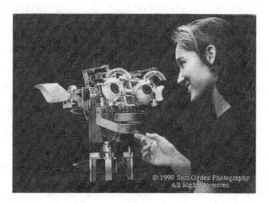

Figure 4.5: An image of Cynthia Brazeal with Kismet (adapted from Breazeal, 1998)

As shown in Figure 4.6, Tokyo University has developed a teacher robot named Saya, which comes pre-programmed with six different emotions of surprise, fear, disgust, anger, happiness, and sadness (Hashimoto et al., 2006). Surely, the ability to deal with affect is well established in robotics, and we will eventually see robots that are emotionally equivalent to humans.

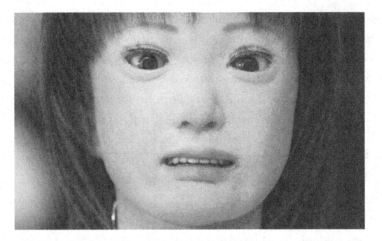

Figure 4.6: Tokyo University's teacher robot (adapted from Takahashi and Takagi (2007))

4.1   CONCLUSIONS

Research on developing affective computing is an active area in robotics (Vallverdu and Casacuberta, 2009). Robotics researchers and practitioners are not arguing for the authenticity of synthetic robot emotions. Instead, researchers are attempting to produce robots in a human world prodigious with emotions. As robots become more present in our daily lives, we must rely on interactions that are rich in emotions. Robotic emotional capacities narrow the existing gap in our quest for robot anthropomorphism. Facial expressions are the main method for communicating emotions. However, non-verbal interactions beyond facial expressions can also convey emotions (Knight, 2011). Movements that convey intentionality often evoke emotions. For example, rapid deceleration or sudden movement freeze could signal an intentional change in a goal. Accompanying feelings could be fear or awe. Even humor can be a useful tool for conveying as well as eliciting emotions (Wendt and Berg, 2009).

Emotions are a step toward robotic anthropomorphism. However, being too human-like is not a desirable property, as is captured by the concept of the *uncanny valley*, coined by Masahiro Mori in 1970. People find robots with very realistic emotions unpleasant and that has plagued humanoid robotics research (Ho et al., 2008). According to Sigmund Freud, the phenomenon that would later be called the uncanny valley stems from a primitive attempt by humans to avoid death and secure our own immortality by creating copies of ourselves—such as wax figures and, later, life-like robots. Our uncanny response follows from the fact that most of us no longer believe we can secure our own immortality by making copies of ourselves, but we haven't yet abandoned the attempt to do so. The sad consequence of this is that, in Freud's words, "The double reverses its aspect. From having been an assurance of immortality, it becomes the uncanny harbinger of death." (Ho et al., 2008). The copies we feel compelled to create only serve to remind us why we began making them in the first place: We are, inevitably, going to die. Throughout his entire career, Mori did not present data to support his proposed uncanny graph. We have no idea whether people will always find humanoid robots "creepy" or if we'll get used to them once they're ubiquitous. 200 years in the future, people may study the theory of the uncanny valley the way people today study unsettling theories about superior and inferior human races from the nineteenth century.

It is self-evident that humans are sentient, self-aware biological machines. It has been argued by the likes of Rodney Brooks that robots are also machines, and they will soon be conscious with well understood mechanisms. Robots will be social, play, entertain, and have a sense of synthetic humor. If robots were to build humans, what would they do with the emotions of their creation?

CHAPTER 5

Sensors

Robotic sensors are electronic devices that a robot uses to collect data from its environment that it deems pertinent for decision making (Everett, 1995). The act of sensing with sensors gathers environmental information. Sensors convert information from a subset of natural phenomena into electrical signals for computational processing. The electromagnetic spectrum component covers light, heat, and radiated signals. The spectrum of molecular motion in the atmosphere covers sound and physical properties such as chemical composition, orientation, and position, thus covering the remaining sources of sensory data. Perception is a robot's ability to become aware of its environment through sensing. Perception fuses and associates sensory data with concepts (percepts) that are meaningful to the robot. While simple perception may choose from a few possible states, complex perception overlaps cognition and goes beyond matching externally gathered sensory data to the most accurate internal, meaningful concept. At times, nontrivial perception is responsible for learning and conceiving of additions to the robot's perceptual repertoire. Sophisticated perception is a key area for future robots.

Perceiving will be responsible for the proper recognition of intention ascribed to surrounding agents. A newly formed, related discipline is *active perception*, heralded by Prof. Ruzena Bajcsy, which explores a continual loop for controlled sensing and perception (Bajcsy, 1995). For the remainder of this chapter, we will limit the scope of our review to the basic sensors available today and refer readers interested in advanced perception (including active perception) to the leading conference series organized by the IEEE professional organization.

The primary focus of this book is more on computational approaches rather than electro-mechanical development. Therefore, this chapter covers the sensing mechanisms that are most popular at the time of writing. Developing sensors is another active area, and complete coverage is not possible. Robots' abilities depend on the quality and speed of their sensors. For robotics, sensing is an *active*, deliberate function. These sensors make observations by emitting energy into the environment to measure reactions or by modifying their environment. In the active sense, they send signals out before collecting reaction data. Active sensors involve direct interaction with the environment and tend to be less energy efficient but more robust in capability. Figure 5.1 shows a sonar sensor emitting an energy wave and receiving the reflected wave. The time elapsed for energy return is a measure of the distance between the robot and objects. Bats use a similar sensor for avoiding hundreds of other bats. In the 1980s, Polaroid cameras (depicted in Figure 5.1) were the most popular application of sonar sensors.

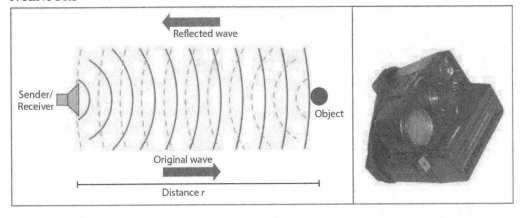

a.   b.

Figure 5.1: Sonar sensor (a) basic Sonar sensor operation (b)Polaroid camera from 1980s

In contrast to active sonar, passive, reflexive sensing in animals (i.e., seeing vs. looking) has been under-explored. A robot's capabilities are directly affected by the speed and quality of its sensors. Passive sensors passively receive energy to make their observations. Passive sensing doesn't emit energy and is especially useful when the robot should be inconspicuous. It generates no noise nor interferes with signals.

Passive infrared sensors (PIR) are used to detect the proximity of humans and animals. Human body heat is radiated at the infrared wavelength. Infrared sensors can, therefore, be used to detect the presence of a person. These systems are popular in both indoor and outdoor security systems and work by detecting the change in infrared thermal heat patterns in front of a sensor that uses a pair of pyro-electric elements that react to changes in temperature. Instantaneous differences in the output of the two elements are detected as movement, specifically movement by a heat-bearing object, such as a human. They consist of lead zirconate titanate plates, which are sensitive to infrared light.

Figure 5.2: A common passive infrared sensor

Sensors vary widely in accuracy and precision. Accuracy refers to the accuracy of a measurement system as the degree of closeness of a quantity to its actual (true) value. Precision in the measurement system (reproducibility or repeatability) is the degree to which repeated measurements under unchanged conditions show the same results.

Sensory information can be fused by combining sensory data or data derived from sensory data from disparate sources such that the resulting information is, in some sense, more meaningful than would be possible when these sources were used individually. This process of selective sensory combination, dubbed *Fusion*, can yield more accurate, more complete, or more dependable results (Waltz et al.,1990). An example is stereoscopic vision, in which depth calculation is performed by combining two-dimensional images from two cameras at slightly different viewpoints (Dasarathy, 1998).

Direct fusion is the fusion of sensor data from a set of heterogeneous or homogeneous sensors (e.g., from a set of sonar sensors in a ring formation, as shown in Figure 5.3), soft sensors, and history values of sensor data. In contrast, indirect fusion uses information sources like a prior knowledge about the environment and human input.

Figure 5.3: A sonar ring that is typically installed in some models of robot base

Far more challenging is fusion using heterogeneous sensory data, such as from vision and infrared.

Range sensors return measurements of distance between the sensors and objects in the environment. Tactile sensors are those used to create a sense of touch. Their general advantages are their simplicity and trustworthiness. Disadvantages are that they often present mechanical problems. Tactile sensors (touch sensors) detect object presence/absence (tactile sensors output a binary value) through physical contact (Figure 5.4a). More sophisticated tactile pads (Figure 5.4b) operate by object imprint shape recognition and, in the future, can be used for improved dexterity of object manipulations.

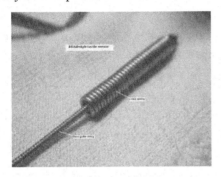
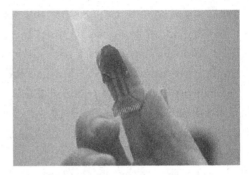

a.   b.

Figure 5.4: Images of typical tactile sensors (a) a simple tactile sensor using two coiled springs and (b) a sophisticated prosthetic hand under development at the University of California Irvine, part of a nationwide effort funded by DARPA

Proximity sensors are non-tactile sensors that are usually active. These include light (infra-red), sound (ultrasound), electromagnetic (capacitance), and magnetic (inductance). Originally invented for space applications, capaciflectors, as shown in Figure 5.5, are capacitance sensors. Electricity creates an electrostatic field that reacts to changes in capacitance. Capacitance sensors are able to sense an object up to 46 centimeters away, and are relatively inexpensive, fast, and light. However, since each object material determines a different dielectric constant, capaciflectors are relatively inaccurate.

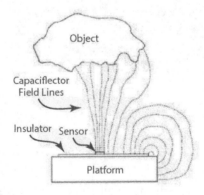

Figure 5.5: A capaciflector

Inductive sensors are electronic proximity sensors, which detect metallic objects without touching them. Shown in Figure 5.6, these easy-to-mount inductive sensors offer powerful resolutions of less than 0.001 mm at sensing distances up to 2 mm.

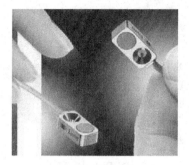

Figure 5.6: Common inductive sensors

Inductive sensors use magnetic fields that induce eddy currents in objects. They are robust and often used in assembly lines. Their disadvantage is that their range is rather small, ranging between 0.0001 in to 1.0 in., making them impractical for many robotic applications.

Ultra-sonic sensors, as shown in Figure 5.3, emit sound waves and measure the sound return time. Their distance measurements are somewhat inaccurate. Factors that affect ultrasonic sensor selection are distance, shapes, and angles. Room corners are particularly problematic due to distortions in signal echoes. Doorways are similarly challenging.

To overcome these problems, their ranges are often overlapped; sometimes they are rotating and bobbing. Although these solutions improve accuracy, they produce crosstalk problems, i.e., interference among emissions and echoes. Perhaps careful timing is the best way to keep their data distinct. Generally, sonar sensors are reliable and inexpensive, but they have poor resolution and often miss some details.

Infrared (IR) sensors emit an infrared pulse and detect the reflected signal. IR is invisible to the human eye. The reflected surface makes a big difference for IR sensors. Their advantage is that they are fast and inexpensive proximity sensors.

Rotation sensors, also called encoders, as shown in Figure 5.7, measure the rotation of a shaft or axle. They are used to measure the angle of a robotic arm or how far a mobile robot's wheel has turned. Encoders as proprioception devices are primarily used for measuring distances traveled.

Figure 5.7: A Vex robot using a shaft encoder (image Courtesy of VEX Robotics, Inc.)

Global Positioning Systems (GPS) receive signals from orbiting satellites that pinpoint the location of an outdoor robot on the Earth. GPS allows the system to determine current location.

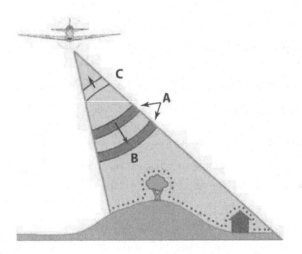

Figure 5.8: A Lidar sensor installed in a airplane

Figure 5.8 shows a basic diagram for a light detection and ranging system (i.e., Lidar). Figure 5.9 shows a Lidar-driven car. Laser range finders use laser beams to measure the distance to objects. They are used for obstacle detection and navigation. Ground penetrating radar (GPR) is a radar-based system; its goal is to detect and identify structures buried in the ground, and it can provide detailed images of the near-earth surface (Daniels et al., 1998). GPR, when applied by the military, operates on the same principle as aircraft radar, in that a controlled pulse of electromagnetic energy is transmitted, reflected off an object, and recorded by a receiver.

Figure 5.9: A Lidar-operated automobile. Credit: Google

The SICK laser scanner, shown in Figure 5.10a, uses mirrors for rotating scanning. This scanner is manufactured by the German company firm SICK. Its range is t*c/2 where t is return time and c is the speed of light (i.e., $3*10^8$ m/sec). They are accurate up to 1.5cm and their range is 1 to 8 meters. They can scan 180° at a 0.5° resolution. The Hokuyo sensor shown in Figure 5.10b is often used to build 3D maps. They are both accurate and reliable. However, these sensors are expensive and heavy.

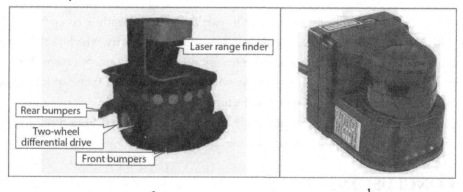

a.   b.

Figure 5.10: A robotic laser sensor (a) a SICK scanner is shown installed on a Pioneer robot. Copyright © 2012 Microsoft Corporation. Used with permission. URL: http://msdn.microsoft.com/en-us/library/bb483042.aspx, and (b) a Hokuyo scanner. Copyright © HOKUYO AUTOMATIC CO., LTD. Used with permission

Visual sensors obtain a considerable amount of data, which can be very expensive to process. Figure 5.11 shows robots with two cameras. *Computer Vision* is the field of study of interpreting camera images for a variety of purposes.

a.   b.

Figure 5.11: Two robots using binocular vision systems (a) the CeDAR active vision head at the Australian National University, from Dankers and Zelinsky (2004), used with permission, and (b) the astronaut assistant robot dubbed foveal extra-vehicular, helper retriever (FEVAHR); adapted from Hexmoor and Bandera (1998)

Occlusion occurs when pixels are missing in parallel images. Depth is computed with f*b / d where f is the focal point of cameras, b is baseline, and d is disparity, which is the image size difference between two cameras. The goal of stereovision is to compute depth maps. Stereovision is accurate for nearby objects. Calibration is the selection of a standard frame of known objects. *Correlation* is the association of corresponding features or points.

Cameras are often offset to produce a 45° cone. Depth is inferred from matching features between images. Typical resolutions of cameras are 240 x 480 pixels at 30 frames/sec.

5.1   CONCLUSIONS

Choosing a sensor involves deliberation over the many characteristics that may be necessary. Physical size, cost, and power requirements are one set of properties to consider. Processing power needed and complementary sensors are a second set of considerations. Clearly, there is a multitude of sensors available, and the many fields of engineering continue to offer us more choices for the vision, auditory, and touch sensory modalities that are critical when perceiving the world. Currently, the senses of taste and smell largely remain unexplored in robotics.

CHAPTER 6

Manipulators

Manipulators (i.e., robot arms) are electromechanical devices that interact with their environment with functionalities that are intended to replicate human arms. Most manipulators are stationary platforms installed on factory floors to handle objects by moving them from one station to another. A robotic arm is a device for grabbing and moving objects by using components that rotate about their ends. The first impression that a robot arm may conjure in common parlance are is the images of a prosthetic arm as shown in Figure 6.1. However, robotics considers a manipulator to be an industrial machine (shown in Figure 6.2) that is designed to move parts on a factory floor (Paul, 1981).

Figure 6.1: Mr. Christian Kandlbauer with a prosthetic arm (image courtesy of Professor Hans Georg Näder and the Otto Bock Group in Austria)

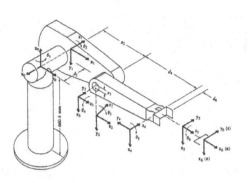
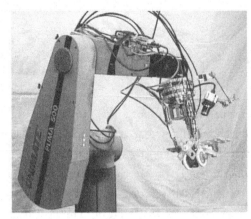

a.   b.

Figure 6.2: A six six-linked robot manipulator called Puma by Unimation, a branch of Westinghouse, inc. (Unimation robotics has been a defunct firm since 1983). (a) A robot schematic for Puma 560 and (b) an image of the Puma 560

A class of manipulators are called *pick and place robots* (see Figure 6.3) since they are programmed to precisely retrieve and place components, such as in an assembly line.

a.   b.

Figure 6.3: Images of pick and place robots. (a) A pick and place robot for circuit board assembly and (b) an image of a robot in a steel cage (i.e., a Gantry robot)

A manipulator is composed of *links* and *joints*. A link is a physical robot component (i.e., a limb). Figure 6.2 shows a Puma 560 robot with six links (i.e., base, shoulder, upper-arm, fore-arm, wrist, and gripper). A joint is a robot component that connects limbs, in which imposes a constraint on the spatial relations between two or more links.

A *ball joint* allows rotation around x, y, and z axes. A *hinge joint* allows rotation around only the z axis, and a *slider joint* (i.e., prismatic joint) permits translation along the x axis. Joints constrain free movements that are measured in degrees of freedom (DOF). Links start with the default 6 DOFs; that is, translations and rotations around three Cartesian coordinate axes. Joints reduce the number of DOFs by constraining some translations or rotations. Often, robots are classified by their total number of DOFs. - (e.g., a Puma 560 (Figure 6.2) is a six six-DOF robot). A *redundant joint* is one that is unnecessary because other joints provide the needed position and/or orientation. A *prismatic joint* is also known as a slider, since the axes of the joint are coincident with the center line of the sliding link. A *revolute joint* is also known as rotary joint, and the axis of the joint is coincident with the center line of the link. A *roll-slide joint* is like a combination knee joint and is commonly found in automobile wheels.

The number of DOF's can be generally determined by *Gruebler's equation* (equation 1).

$$F = 3 \times (n-1) - 2 \times j - 1 \times h \quad (1)$$

F = the number of DOF's

n = the number of links, including ground

j = the number of revolute plus prismatic joints

h = the number of roll-slide joints

A *passive* degree of freedom allows an intermediate link to rotate freely about an axis defined by two joints. A passive DOF cannot transfer torque. Mechanically, arm links can be seen as *kinematics chains*. The chain is closed when the ground link begins and ends the chain; , otherwise, it is an open chain (Spong and Vidyasagar, 1989). A *parallel robot* is a closed closed-loop chain, whereas a *serial robot* is an open open-loop chain (e.g., the Puma 560).

In order to find a transformation from tool tip (i.e., the outermost end of the manipulator) to the base of the manipulator, we have to define link frames (i.e., a Cartesian coordinate system) and derive a systematical technique, which allows describing the kinematics of a robot with n degrees of freedom in a unique way. A set of four parameters is proven sufficient for this purpose. According to Denavit-Hartenberg (1955) notations, only four parameters (a, d, q, a) are necessary to define a frame in space (or joint axis) relative to a reference frame (see Figure 6.4). Figure 6.5 shows a view of all points that are reachable.

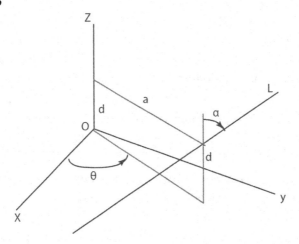

Figure 6.4: The Devavit and Hartenberg notation

a = minimum distance between line L (the z axis of the next frame) and z axis
    (mutually- orthogonal line between line L and the z axis)
d = distance along z axis from z origin to minimum distance intersection point
q = angle between x-z plane and plane containing z axis and minimum distance line
a = angle between the z axis and line L

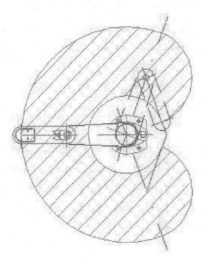

Figure 6.5: Top view of a robot work envelope

Kinematics describes the analytical relationship between the joint positions and the end-effector position and orientation. *Differential kinematics* describes the analytical relationship between

the joint motion and the end-effector motion in terms of velocities. There is a kinematic relationship between any pair of frames: a translation and a rotation. This relationship is represented by a 4 × 4 homogeneous transformation matrix. There are two basic kinematic tasks. First, direct kinematics (i.e., forward); that is, . Aassume we are given joint relations (i.e., rotations and translations) for the robot arm. The question that needs an answer is, "What is the orientation and position of the end effector?" Second is inverse kinematics (i.e., backward); that is, . Aassume that we are given the desired end effector position and orientation of the end effector. The question that needs an answer is, "What are the joint rotations and orientations to achieve this?"

A chain of joints is computed by matrix multiplication: $X_{n-1} = A_{n-1} A_{n-2} \times .. A_1 \times A_0 \times X_0$. Here, X_i is the point in the ith coordinate frame, and A_i is the ith transformation matrix. Keep in mind that $X_i = A \times X_{i-1}$. For a kinematic mechanism, the *inverse kinematic problem* is when given an end point, all joint angles need to be computed to support a valid robot configuration, which is a difficult problem to solve. The robot controller must solve a set of non-linear simultaneous algebraic equations.

The Selective Compliant Assembly Robot Arm (SCARA) is a 1981 NEC standard in which the arm is slightly compliant in the X-Y direction but rigid in the Z direction. SCARA robots are generally faster and cleaner than comparable Cartesian systems. Figure 6.6 shows an Epson SCARA robot.

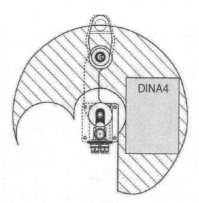

Figure 6.6: An Epson SCARA robot. A four axis SCARA Epson G3 robot with max payload of 3 kg, and precision of ±0.01mm (image courtesy of Epson.com)

Figure 6.6 shows *work envelopes* that are loci of all points that the robot can reach and perform work as surrounding spaces of robots, in which the robots are able to reach. Figure 6.5 showed a aerial view of work envelope. Figure 6.6 shows a SCARA robot and its work envelope.

Figure 6.7 shows an end effector that is a simple set of pair of claws acting that act like an object gripper. More sophisticated than grippers, robotic hands exist that are far more dexterous and complex than simple end effectors. Figure 6.8 shows an image of a Utah/MIT hand designed in the mid-1980s. It consists of four 4-DOF modular digits driven by an impressive set of artificial tendons and pulleys.

Figure 6.7: A simple gripper

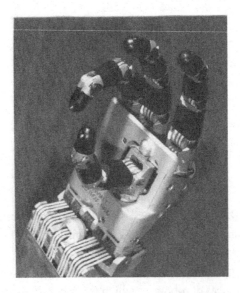

Figure 6.8: The dexterous Utah/MIT robot hand (image adapted from Jacobsen, et. al. et al., 2003)

There is a universal gripper developed at Cornell that is a party balloon filled with coffee grounds. By using vacuum power, it sucks up objects with odd shapes and sizes, such as a coffee mug (Brown, et. al. et al., 2010).

Teach pendants, as shown in Figure 6.9, are hand-held devices that an operator/technician uses to record a series of successive points and poses for an arm to visit. This interface device reduces robotic programming to mimicking prerecorded points. In a robot-equipped factory work cell, typically there are thousands of routes and patterns stored in memories and function libraries. Manipulators are more effective when they are combined with sensing systems, such as vision systems. They can also be installed on mobile platforms for greater reach.

Figure 6.9: A teach pendant. Image courtesy of Comau (http://www.comau.com)

The Comau teach pendant (Figure 6.9) is used to program various models of the robotic industrial machinery that the company makes.

6.1 CONCLUSIONS

Robotic manipulators have revolutionized manufacturing assembly assembly-line productivity and perform a multitude of important dexterous, repetitive operations that require high levels of precision, including drilling, welding, transporting, fastening, and packaging. Popular culture has promulgated the cautionary negative stereotype that robots are replacing human factory workers. In the West, we've had over a century to adjust to automation, and the common way we've coped is to divert to a more service-based economy. Retraining and retooling the work force is essential because robotic automation is here to stay. In the future, we will see more varied and versatile forms of robotic arms.

PART II

Mobility

Part I of this book provided a rudimentary, broad introduction to robotics. At this point we turn to the fundamental techniques and various approaches to mobility, which includes decisions about location and navigation in the environment.

CHAPTER 7

Potential Fields

The farthest reach (i.e., the tip) of a manipulator or an entire mobile robot can be considered to move in a field of forces, creating a *potential field* (PF). The desired positions to be reached are treated as attractor poles while obstacles are treated as repulsive surfaces (repulsive poles) for the robot. Potential fields, also known as *vector fields* in mathematics, reflect a snapshot of the current orientation for a cluster of moving objects at a given moment. An electric potential field, this concept is related to the potential energy of a positive test charge at various locations within an electric field. PFs cause the robot to appear to "slide around" obstacles. PF is defined as the sum of the attracting and repulsing forces in the field. Attracting forces get smaller as the robot approaches its goal. Repelling forces get larger as the robot approaches obstacles. PF is often constructed with a fixed radius of effective forces. For a detailed, mathematical, and algorithmic coverage of potential field methods, the reader may consult Latombe's *Robot Motion Planning* (Latombe, 1991).

In physics, objects may possess a property known as an *electric charge*. An *electric potential field* exerts a force on charged objects, pushing them in the direction of the force, which is in either the same or the opposite direction of the electric field (Harper and Weaire, 1985). If the charged object has a positive charge, the force and acceleration will be in the direction of the field. This force has the same direction as the electric field vector, and its magnitude is the size of the charge multiplied by the magnitude of the electric field. Classical mechanics explores concepts such as force, energy, potential, etc. in much more detail.

While a target (i.e., a goal) exerts an attractive force, obstacles produce repulsive forces. Magnitude and directions of these forces are combined to determine an overall navigation direction for the robot (see Figure 7.1). Figure 7.2 shows an example of a potential field with one obstacle and one goal. A rotational potential field is shown in Figure 7.3. For each point (x, y), an attractor or a repulsor exerts a corrective vector that we'll denote by deltas force or a gradient of a potential function P (x, y), as shown in Figure 7.1.

a.
$$(x, y) \rightarrow (\Delta x, \Delta y)$$

b.
$$(\Delta x, \Delta y) \quad = \quad \nabla P(x, y) \quad = \quad \left(\frac{\partial P}{\partial x}, \frac{\partial P}{\partial y} \right)$$

Figure 7.1: Potential vectors (a) corrective force vector for a point (b) the gradient of a potential function $P(x,y)$

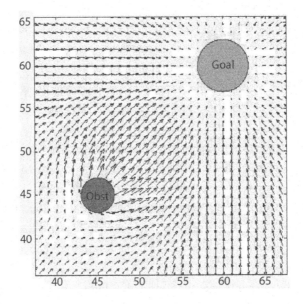

Figure 7.2: An artificial potential field (adapted from Goodrich, 2000)

The most well-known problem for using potential fields is the box canyon problem shown in Figure 7.4, in which the robot is attracted to a goal beyond a wall. As the robot enters the canyon zone, it is repelled by canyon walls but soon reaches the canyon's end. Blocked by the wall, it cannot reach the goal. Another problem is local minima, created by conflicting forces. Figure 7.5 shows an attracting goal flanked by two repulsive obstacles with forces that add up to zero and create a local minima problem. The robot moves toward the goal, but is repulsed such that it cannot travel between obstacles to reach it.

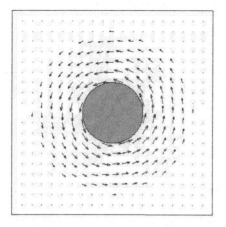

Figure 7.3: A rotational potential field (adapted from Goodrich, 2000)

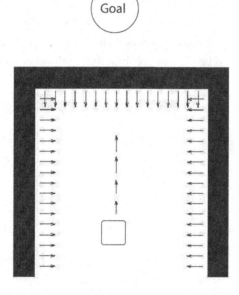

Figure 7.4: The box canyon problem (adapted from Goodrich, 2000)

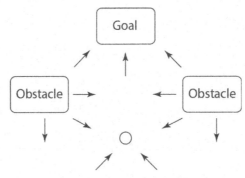

Figure 7.5: Local Minima created by conflicting objects

7.1   CONCLUSIONS

Artificial potential fields have been one of the oldest, simplest methods used for robotic navigation (Dunias, 1996). While often elusive, if the environment can be simplified to attractive and repulsive poles, navigation can be reduced to a set of vectors that steers the robot through its environment. There might be problems with opposing forces that cancel each other out and leave the robot stranded. Therefore, PFs need to be used in combination with other techniques to avoid pitfalls.

CHAPTER 8

Roadmaps

The coming age of driverless cars is on the verge of revolutionizing the automobile market. In the next 25 years, driverless cars and global positioning systems will replace human-piloted cars. A *roadmap* is a topological map with places of interest and a set of paths or roads between those places, were relationships among places are maintained. A *road* is an obstacle-free path between two points in the environment.

Roadmaps may lack scale and distance, and directions are subject to change and variation (e.g., the Chicago street map or the London tube routes illustrate relationships among a set of places of interest). Topology is the branch of mathematics that studies the properties of objects that do not change as the object is deformed (Munkres, 2000). For a more detailed, algorithmic coverage of roadmap methods, consult Latombe's *Robot Motion Planning* (1991).

Roadmaps are typically represented in a graph structure, with nodes for places and edges as paths. Roadmaps are most commonly used as strategies for finding pre-planned paths in a robot environment. We assume that the robot has access to the entire map prior to navigation (i.e., the robot possesses global knowledge of the environment).

The roadmap graphs are usually pre-computed without knowledge of starting and ending locations. Graphs are stored as *declarative knowledge* (in contrast to *procedural knowledge*) as in Prolog programs. Specific start and stop locations are added later, via queries to the map, sometimes requiring additional edges to be added. In every case, *path discovery* is a map search to find the shortest, the fastest, or the most efficient path (Govindan and Tangmunarunkit, 2000).

A *path* is the result of a query to the map database that has two end points. The map is searched for an efficient (i.e., shortest) distance from the start point to the finish point. The most common search algorithm might be Dijkstra or variations of A* algorithms (Kleinberg, 2005).

There are two basic ways to store maps. A *vector representation* uses a graph of nodes and edges to denote vertices of polygons that approximate obstacles in the environment. For an environment with relatively few objects, vector representation is efficient. In contrast, a *raster representation* divides the environment into small cells (similar to the pixels of an image) and records free versus occupied cells in the environment (also called *occupancy grids*). An environment with more complex objects, or just more objects, can be more accurately represented with raster representation. Computationally, vector representations are more tractable but record rather coarse information about objects. Although raster representations are richer (i.e., inversely proportional to cell sizes), they are more computationally challenging.

Broadly speaking, there are two types of road map analyses. *Geometry-based algorithms* use mathematical geometric models of polyhedral obstacles and compute nodes and edges based on given constraints. In contrast, *sampling-based models* create the graph from randomly selected, interconnecting robot configurations in the configuration space.

8.1   GEOMETRY-BASED ROADMAPS

Geometry-based roadmaps use vector representations where obstacles are approximated as concave polygons, with possible loss of obstacle detail. Robot start and end points are marked.

From the starting point, the robot orients itself toward its destination and builds a *visibility graph* (VG). Successive lines of sight from those points extend the map until the endpoint is visible. VGs are non-directed graphs used to discover goals and avoid obstacles. VGs typically address a set of points and obstacles in the Euclidean plane (Lozano-Perez and Wesley, 1979). Each node in the VG graph represents a place denoted as a point, and each edge represents a visible connection between a pair of points. If the line segment connecting two points does not intersect any obstacle, an edge is drawn between them in the graph. In a visibility graph, the nodes are mapped to the vertices of polyhedral convex obstacles, as well as to the start and the destination points. Likewise, viable edges are comprised of any and all connections between nodes, which do not transit through an obstacle. These viable edges, as finite segments of visibility rays, are referred to as *support lines* and are valid paths of traversal. There is an obstacle to the right and to the left of each support line. It is important to remember that the edges of obstacles may also be considered as support lines and that visibility graphs can only be used in environments with convex polyhedral obstacles. If a support line touches vertices of an obstacle more than once, that obstacle is concave. If obstacles are not already convex polyhedrons, we can convert them, but this is beyond the scope of this book. Consult Wenninger's *Polyhedron Models* (1974) for more details on geometrical conversion.

Steps for building a visibility graph (see Figures 8.1–8.7):

1. Perform a *radial sweep* to pick out support lines (including the destination). A radial sweep emits a ray from the starting point and continually, radially moves the ray, centered at the start. Meanwhile, note the places where the ray tangentially contacts obstacles. Similar to a traditional radar screen sweep, the sweep identifies obstacle extremities (i.e., vertices). Figure 8.1 shows the ray touching obstacle vertices.

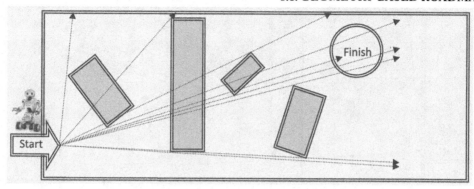

Figure 8.1: The radial sweep with four obstacles

2. Remove all invalid paths by eliminating any segments that intersect an obstacle. Figure 8.2 shows a few lines removed from the Figure 8.1 radial sweep.

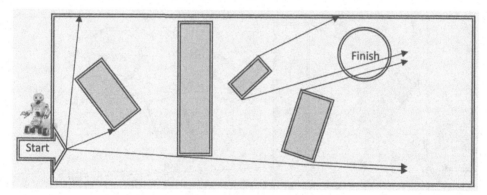

Figure 8.2: Eliminate obstructed paths

3. Repeat iteratively using each obstacle vertex (node) as a new starting point. Figure 8.3 shows an application of this process.

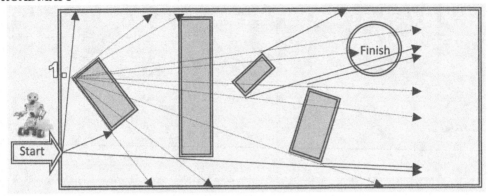

Figure 8.3: Move to next node and sweep again

A radial sweep is repeated at all obstacles vertices as exemplars, as shown in Figures 8.4, 8.5 and 8.6. Once all sweeps are completed, the line segments are compiled into a graph, as shown in Figure 8.7.

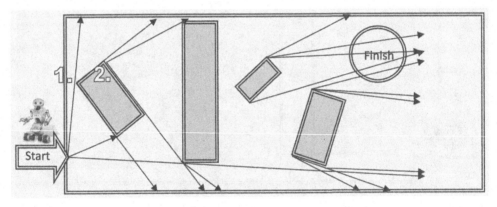

Figure 8.4: Move to the next node and sweep again

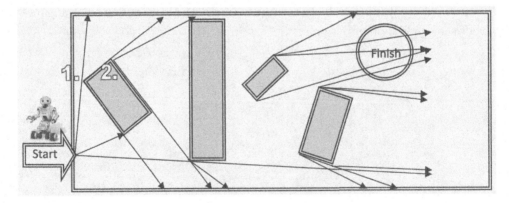

Figure 8.5: Remember to remove obstructed paths

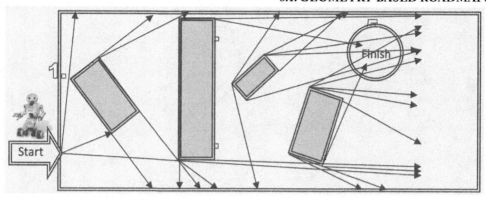

Figure 8.6: Repeat steps 3 and 4 exhaustively

4. Append all support lines to create a visibility graph. The graph can be created in O (n^2 log n) running time.

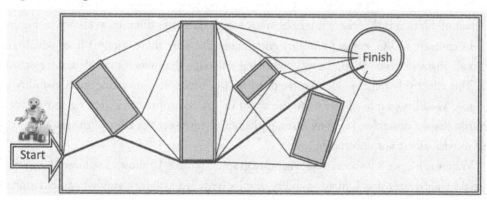

Figure 8.7: The completed visibility graph with shortest path

The shortest path from the start to the destination can then be found using a search through this graph. Several search options are available, and the most popular ones are listed here:

1. **Dijkstra's algorithm:** This is a simple algorithm used to find the shortest path in a graph (Cormen et al., 2001).

2. **A* algorithm:** This extension of Dijkstra's algorithm utilizes a best-first search to find the shortest path in the graph. A distance-plus-cost heuristic function determines the path of the search (Cormen et al., 2001).

An important point to remember is that the actual robots are not points and have size; this is an added dimension that will collide with obstacles. Figure 8.8 demonstrates the idea of dilation.

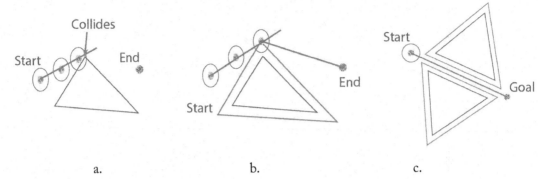

Figure 8.8: The idea of obstacle dilation (a) robot collides with obstacle without dilation, (b) robot avoids obstacle that is dilated by robot diameter, (c) problem with narrow passages after dilation

To prevent problems with physical robot size, the obstacles are dilated, as shown in Figure 8.8 a and b; i.e., "fattened" or "grown" to create a safety buffer (Rawlson and Jarvis, 2007). A negative side effect of dilation is that there might be narrow passages after dilation, as shown in Figure 8.8c.

In contrast to VG, *reduced visibility graphs* consider only those support lines which are tangent to all obstacles to be valid, thereby removing any edge that, when extended, intersects an obstacle. This effectively removes all concave portions of obstacles. Converting from visibility graphs to reduced visibility graphs allows us to use them in any environment, including those containing arbitrarily shaped obstacles. In many cases, paths which intersect the edge of the map can be eliminated as well, albeit not universally.

Whereas Figure 8.9 shows a full visibility graph, Figure 8.10 shows a reduced visibility graph. Figure 8.11 illustrates that limited visibility graphs (reduced visibility graphs) can circumnavigate concave obstacles.

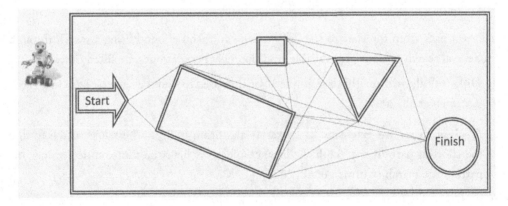

Figure 8.9: A full visibility graph containing 29 edges

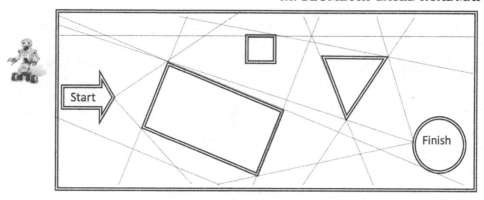

Figure 8.10: A limited visibility graph containing 13 edges

Figure 8.11: Limited visibility graph's ability to map concave shapes: 9 edges

8.2 GRIDS

Another approach to robot path finding involves navigating an occupancy grid. Cells on the grid are given weights based upon their distance from obstacles. Values are highest at closest distances to obstacles and drop off proportionally to distances away from them. Cell values are processed according to the distance from the source, and this creates a *wave front* or *brush fire* effect. For more details on these effects, consult Dudek and Jenkins' Computational Principles of Mobile Robotics (2010).

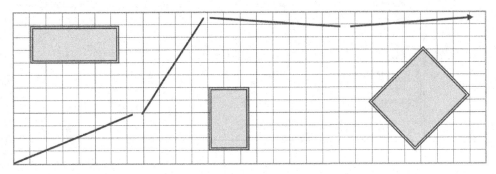

Figure 8.12: An occupancy grid with obstacles and a path through it

By following the cells with lowest weights, a robot can traverse a pre-mapped terrain as shown in Figure 8.12. This technique can also be used to insure that non-point robots are able to avoid running into obstacles by specifying the upper limit of the value of the cells they can occupy. This is reminiscent of the navigation with artificial potential fields outlined in Chapter 7. Alternatively, an occupancy grid can be converted into vector obstacles that can apply the vector–based algorithms.

Obviously, resolution is an issue when using grids to find paths. Large grids have a lower resolution and may result in blocking paths that can be visible by using finer-grained grids. One important issue to remember is that the minimum size of each region must be smaller than the smallest gap between obstacles.

Voronoi road maps are a variant of the grid-based approach and provide a technique to insure paths are created that offer maximum clearance between obstacles. This technique is popular because it decreases the reliance on sensors, which can be very noisy, inaccurate, and prone to error. A voronoi diagram is a map partitioned into voronoi cells. The definition of a voronoi cell is "a set of all points that are closer to a given point, S, than any other given point." If each obstacle is specified as a given point, the divisions between the given cells would represent the possible pathway of the robot. Figure 8.13 shows an example of an environment divided into voronoi cells (i.e., a voronoi diagram).

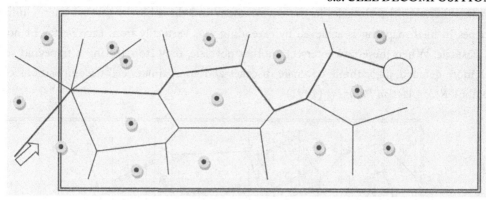

Figure 8.13: A voronoi diagram from points

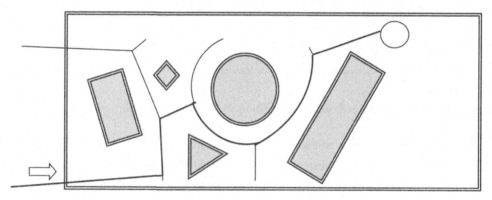

Figure 8.14: A generalized voronoi diagram from polygonal obstacles

If, as in the example just given, the obstacles are not points but polygons, it is necessary to compute a *generalized voronoi diagram* (GVD) with the edges of each cell maintaining maximum distance from both the obstacles and the edge of the environment. This can be done by assigning given points the entire length of the sides of the polygons (Gavrilova, 2008). Figure 8.14 shows a GVD diagram (Siegwart and Nourbaksh, 2004).

A voronoi diagram can also be created using the technique with the grid. By continually increasing the value of cells in a grid that are adjacent to obstacles, the result will be a diagram in which the cells with the lowest values are necessarily those farthest from all obstacles. Once possible paths are computed the shortest, quickest, or most efficient path can be found.

8.3   CELL DECOMPOSITION

There are many ways to segment a given environment into cells. In each case though, the end result is always relatively small polygonal regions. One simple way to divide an environment is to create trapezoids and triangles by using vertical dividers, as shown in Figure 8.15. Free space becomes segmented between the obstacles in the environment and the arbitrary vertical lines. Uniformity of shapes in final mapping is achieved by extending rays vertically from each vertex or node on each obstacle. When these rays intersect another obstacle, they stop, leaving a trapezoid section. For a more detailed, algorithmic coverage of exact and approximate cell decompositions consult Latombe's *Robot Motion Planning* (1991).

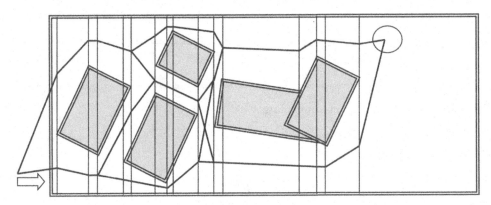

Figure 8.15: A vertically divided cell decomposition with possible midpoints paths marked

By finding the midpoint of each of the vertical dividers and then connecting them, a road map maximizing distances from obstacles is created. By further computing the start and end points by determining in which trapezoid they lie, an easily searched graph becomes evident.

This roadmap can be further optimized by adding additional points to the vertical lines. The increased choice in paths leads to more processing time, but also the possibility of a more efficient path. One risk when doing this, however, is that the robot may move too close to an obstacle. Care must be exercised to insure the points are not placed too close to an edge.

Boustrophedon cell decomposition is a variant of the above technique that utilizes only those nodes from which a vertical line can be extended both up and down through free space, as shown in Figure 8.16. The cells created by this technique are, as a rule, no longer trapezoids. This leads to fewer cells but more complex cell shapes. This subdivision of the environment is used in various strategies, making the problem size much smaller.

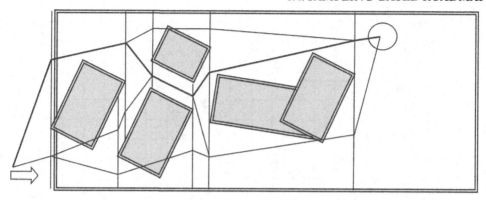

Figure 8.16: A boustrophedon cell decomposition with the only possible midpoints path marked; all others have been disqualified due to crossing obstacles

Another approach to the decomposition problem is *Canny's Silhouette* method (Choset et al., 2005). This strategy subdivides the environment into silhouette curves, each representing the border of an obstacle. This can be envisioned as a vertical line moving horizontally across the environment. As it moves, it will be subdivided by the same points used in boustrophedon cell decomposition. Likewise, at the end of the obstacle, lines will merge at the same points again. In this case, however, the path is mapped to the extreme points of the vertical line as it passes through the start and on to the end.

8.4   SAMPLING-BASED ROADMAPS

These roadmaps are characterized by connecting fixed or random valid robot positions. These connections are usually based on close proximity. A valid robot position is a *robot configuration*, which is a specification of a robot position and orientation relative to the robot environment's reference frame (Latombe, 1991).

Grid-based sampling is perhaps the simplest method. Grids are overlaid to cover "free" spaces. As previously explored in section 8.2, grid size does matter, so picking a grid sized smaller than the obstacle separation distance is required. Figures 8.17 and 8.18 show 4- and 8-connectivity maps, respectively, for an environment with five obstacles. Similarly, Figures 8.19 and 8.20 show another environment with the same comparison.

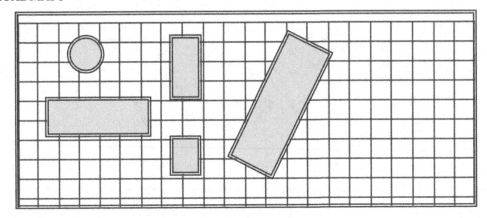

Figure 8.17: A map with a 4-connectivity grid overlay on obstacles

One aspect of grids that has not been discussed is the shape of standard cells. A grid could be connected via a 4-connectivity grid, making a pattern of many squares in the free space, or each square could be horizontally bisected, making four triangles via 8-connectivity. The more inter-connected a grid is, the more efficient a path can be found, at the expense of increased space and computation time.

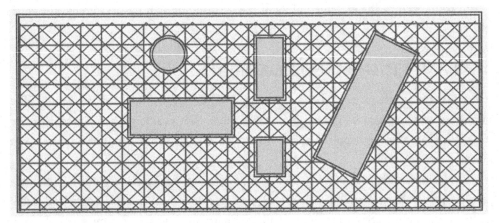

Figure 8.18: An 8-connectivity grid on the map used in Figure 8.17

Hierarchical maps are possible by utilizing a course grid and moving to finer and/or more interconnected grids along a rough path. Figure 8.19 is an example of a hierarchical map.

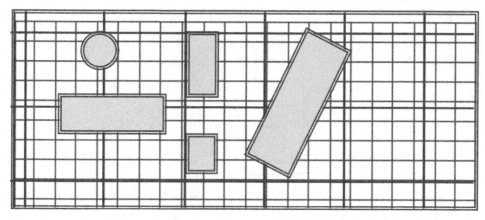

Figure 8.19: A 4-connectivity hierarchal map

8.4.1 PROBABILISTIC ROADMAPS (PRM)

Probabilistic Roadmaps are created by randomly selecting free points and interconnecting them, creating a graph. Figure 8.20 is the pseudo code for generating a PRM graph: G = (V, E). Figure 8.21 gives us an example using a pictorial illustration.

1. **Let V** and **E** be empty
2. **Repeat**
    a. **Let v** be a random robot configuration (i.e., random point)
    b. **IF** (v is a valid configuration) **THEN** add **v** to **V**
3. **Until V** has **n** vertices // from step 2
4. **For** (each vertex **v** of **V**) **DO**
    a. Let **C** be the **k** closest neighbors of **v**
    b. **For** (each neighbor c_i in **C**) **DO**
       IF (E does not have edge from **V** to c_i) and (path from **v** to c_i is valid) **THEN** add an edge from **v** to c_i to **E**
    c. **END** // from step 4b
5. **END** // from step 4

Figure 8.20: PRM algorithm

The difficulty in using PRM is that of identifying the neighbors to which any node should be connected. Several strategies exist to accomplish this:

- **Brute Force:** Entails exhaustive interconnections followed by the discarding of paths that intersect obstacles; what remains are the possible paths.

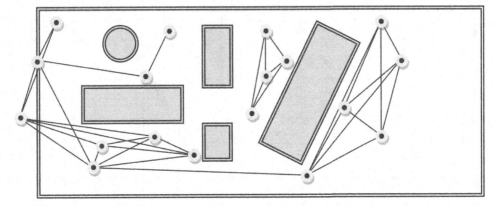

Figure 8.21: A map showing brute force approach to Probabilistic Roadmaps

Note that in Figure 8.21 all connections crossing obstacles have been removed, leaving an unconnected graph, which is one of the risks associated with probabilistic roadmaps.

- KD Trees: The most popular method of building probabilistic maps. They are constructed by alternately subdividing cells (and points within those cells) with vertical and horizontal lines. The final goal of this division is to have one point per cell. A binary tree can be created with points as terminal nodes and adjacent points as children of the same parent (based upon the final subdivision of each cell).

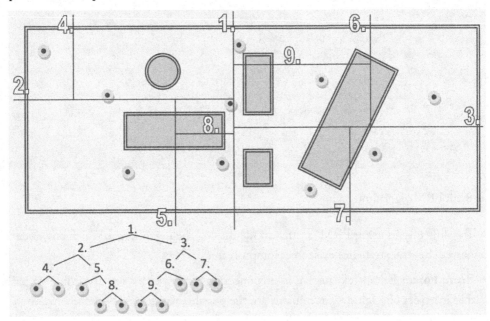

Figure 8.22: A map illustrating KD Tree with associated map. The tree is then used to identify probable neighbors

- R-Trees
- VP-Trees

Probabilistic Roadmaps perform well on average, but, each case has a chance of having no solution. Due to the random placement of the points and connection paradigm used, disconnected graphs can result (as illustrated in Figure 8.22).

8.4.2 RAPIDLY EXPLORING RANDOM TREE (RRT)

Rapid exploring random trees are built by constructing a graph rooted at the start location and randomly growing toward the goal. Each node in the tree represents a valid possible robot location as well as a "branching" point. The direction of movement from each node could be biased toward the goal, or it could be random. Figure 8.23 offers the RRT algorithm for G = (V, E).

> 1. **Let V** contain the start vertex and let **E** be empty
> 2. **Repeat**
>   a. **Let q** be a random valid robot configuration (i.e., a random point)
>   b. **Let v** be the node of **V** that is closest to **q**
>   c. **Let p** be the point along the ray from **v** to **q** that is at distance s from **v**
>   d. **IF** (**vp** is a valid edge) **THEN** add new node **p** to **V** with parent **v**
> 3. **Until V** has **n** vertices //

Figure 8.23: RRT algorithm. Parameters in the algorithm:
- Number of nodes represented by n
- Edge Length (step size) of exploration represented by s

RRTs have difficulties with tasks such as navigating narrow passages and navigating around obstacles, especially concave ones.

There are many alternatives of RRT. We can bias the results to selectively head toward the goal at some percentage of time. We can guide RRT by using the goal-point for expansion direction instead of random configuration points. If we wish to rapidly reach the goal, we may prefer the goal for expansion at a high 95% rate. If we select the goal at a low 5% rate, we'll approach it at a leisurely, slow pace. If percentage of goal selection is small, it takes longer to reach the goal. If the percentage is large, a robot will progress faster, but it may get stuck in a local minimum. The *greedy approach* to RRT growth is to allow the tree to expand beyond the stepping distance of s. If step is kept small, it takes long to reach the goal. If the step is large, it will expand faster, but it may overstep the goal.

8.4.3   DUAL TREE

Once a complex environment is triangulated, a dual graph can be constructed. Triangulation is the breaking of the environment into triangles. To construct a Dual Graph (DG), perform triangulation. Next, place a vertex at the center of each triangle. Add an edge between vertices only if they share a triangle edge. Vertices and edges of a DG are used as input to path-finding algorithms (e.g. Dijkstra's Shortest Path and A*). A DG can be used on its own for path finding. Build two trees, one rooted at start the other rooted at end. As they grow together, they meet in the middle. By merging the trees, paths can be found and a search for the most efficient can begin. Thin triangles can create dangerous paths since the paths may clip obstacles (Rao, 1995).

8.5   OBSERVATIONS

The roadmap methods discussed in this chapter are used in research laboratories as well as in the real world. For example, bowling-ball sized free-flying satellites called Synchronized Position Hold, Engage, Reorient, Experimental Satellites (SPHERES) are designed to be controlled by smartphones (Figure 8.24). Each SPHERE is self-contained with power, propulsion, and computational and navigational equipment. One of the roadmap methods proposed for SPHERES is a template route planner (Slack, 1993), which was shown to be effective on a mobile robot in the Saltonsea desert in California (Miller et al., 2003).

Figure 8.24: Expedition 22 Commander Jeff Williams performs a check of the SPHERES Beacon/Beacon Tester aboard the International Space Station (image credit: NASA)

8.6   CONCLUSIONS

Roadmaps have been one of the most extensively explored areas of robotics. The need for outdoor map building disappears if we assume access to a reliable global positioning system. There are a few ongoing commercial projects at Google and elsewhere that offer indoor maps for public use. However, in absence of available maps, robots need to build and use their own maps. This chapter reviewed rudimentary concepts for storing, building, and using maps, best covered in Latombe's *Robot Motion Planning*. Although algorithmic improvements are continually reported, computational map-processing speed is the main limiting factor. To a lesser extent, more reliable and inexpensive proximity sensors will advance the research and development of roadmaps.

CHAPTER 9

Reactive Navigation

Robot navigation is the behavior of moving a robot from one location to another in a collision-free path. *Reactive navigation* is the study of techniques for planning, recording, and controlling the course and position of a robot toward a desired location, where knowledge of objects in the environment is only known after the movement begins. Navigation is a set of reaction strategies to successfully maneuver around obstacles, avoiding collision while minimizing travel distances. This has been likened to an insect's simple-minded movement to the food source, often called *bug algorithms* (Choset et al., 2005). We will examine algorithms that work in environments with known, as well as unknown, obstacle locations.

The goal is to understand various ways of representing maps. This will include investigation of a way to navigate using *feature-based maps*. Alternatively, we may also investigate a way to navigate using potential fields. We will determine how a robot can make the "best" decision based on only local sensor information.

During navigation, robots either move toward their desired target location(s), or follow a fixed path that is known in advance. When heading toward a destination, the robot usually relies on local sensor information and updates its location and direction according to the best choice that will lead to the goal. When a fixed path is provided on which to navigate, the path is usually computed (planned) prior to navigation (Latombe, 1991).

Path planning is the behavior of examining known information about the environment and computing a path that satisfies one or more conditions such as (a) avoid obstacles, (b) travel the shortest distance, (c) make the least number of turns, and (d) travel the safest paths.

The main objective for path planning is efficiency. In real robots, the optimal solution is not always practical. Instead, approximate solutions are often sufficient and preferred. To perform any nontrivial movement, a mobile robot must pre-plan its paths.

Many interesting problems are solved that make use of the planned motion of the robot. Global navigation strategies for mobile robots can usually be obtained by searching for a collision-free path in a two-dimensional floorplan of the environment (Lozano-Perez, 1983).

We will look first at *goal-directed navigation*, in which the robot is trying to reach a predetermined goal location. If the robot and goal locations (their coordinates) are given, goal position would be given as a point in the coordinate system, and navigation is reduced to a problem where the robot must maintain its own location as it moves (e.g., dead reckoning, which is inaccurate), or have this information collected externally (e.g., using a GPS system or a set of sensors for localization).

If the locations and shape of obstacles are known, we'll assume that coordinates of all polygonal obstacle vertices would be given and known to the robot. Otherwise, the robot must be able to sense obstacles, and sensing is prone to errors and inaccuracies.

If the locations of the destination as well as all obstacles, are unknown, the robot will rely on its innate behaviors such as wandering, wall- or line-following, or any of the simple Braitenberg vehicle behaviors discussed in Chapter 2 (e.g., light seeking).

Persians, Arabs, and Indian Ocean islanders such as the Maldives traversed the open seas with reactive navigation (Lewis, 1970). If the location of a destination is unknown, whereas obstacles are known, navigation becomes either a *target search problem* or an *area coverage problem* that will be discussed in section 9.8.

If the location of the destination is known, whereas shapes and distances among obstacles are unknown, the robot must be ready to produce reactive behaviors for dealing with obstacles, known as *reactive navigation problem*, which is the main focus of sections 9.1–9.7 of this chapter. We assume that the robot's capacity for discovering obstacles is strictly limited to local sensing.

If the locations of the destination as well as all obstacles are known, the robot possesses global information about its environment. We'll treat this scenario with navigation methods that are based on features or potential fields. Alternatively, we offer roadmap-based planning.

In the next section, we will consider solutions for the reactive navigation problem. A robotic application that motivates navigation under our assumptions is *close-range inspection*. Consider scenarios where the robot is surveying a particular area. When it finds an object of interest, it will get closer to the object to obtain more details. Thus, the robot will require a navigation strategy to converge on the target object in an environment cluttered with many objects. There are many applications areas, including military, medical surgery, and construction. For example, the WedgeBug robot was proposed for a sojourner rover intended for Mars exploration.

9.1   BUG ALGORITHMS

We assume that the environment may continually change and little information about the environment may be completely reliable at any given time. The navigation strategy must reach the target with as little information as possible, given only the current robot position and the target location (Lumelsky and Stepanov, 1987).

To simplify presentation, we adopt a few assumptions to approximate the robot to a bug. First, the robot is a point object; i.e., the robot has no size. Second, the robot has reliable localization capability; i.e., the robot knows its true position and orientation relative to the origin at all times. Third, the robot has reliable and accurate sensors; i.e., the robot will use its distance sensors for navigation.

9.2   BUG 0 STRATEGY

Let us assume that the robot knows the goal location, but is unable to detect it since it lies beyond at least one obstacle. Bug 0 strategy consists of the three simple steps shown in Figure 9.1 (Choset et al., 2005). Figure 9.2 shows an example, as follows. The robot moves toward the goal. When it discovers an obstacle, it suspends seeking the goal and deals with moving around the obstacle. Once the obstacle is cleared, the robot returns to goal-seeking behavior. Here we assume a left-turning robot. The turning direction is decided beforehand, i.e., the goal -seeking behavior is subsumed by the behavior for obstacle avoidance.

1. Head toward the goal
2. Follow obstacles until you can head toward the goal again
3. Continue

Figure 9.1: Bug 0 algorithm

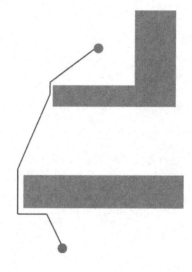

Figure 9.2: Bug 0 (adapted from Choset et al., 2005)

9.3   BUG 1 STRATEGY

This basic strategy is to move toward the goal until an obstacle is encountered, then go around the obstacle in order to find the closest point from the perimeter points of the obstacle to depart toward the goal. If needed, the robot travels back to that closest departure point and resumes moving toward the goal. Successive obstacles after the first are similarly handled. Figure 9.3 is an example and Figure 9.4 outlines the algorithm.

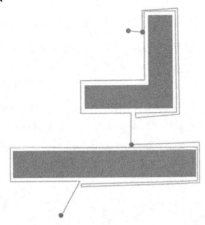

Figure 9.3: An example illustrating Bug 1 algorithm (adapted from Choset et al. 2005)

1. WHILE (1)
   r is the robot's location; g is the goal location
2. REPEAT
   Move from r toward g
3. **UNTIL** ((r==g) **OR** (obstacle Is encountered))
4. IF (r==g) THEN quit // goal reached
5. **LET** p = r // contact location on obstacle
6. **LET** m=r // obstacle perimeter location closest to g so far
7. **REPEAT**
   a. Follow obstacle boundary
   b. r = robot's current location
8. IF ((distance(r,g) <distance(m,g)) **THEN** m=r //update m
9. **UNTIL** ((r==g) OR (r==p))
10. **IF** (r==g) **THEN** quit // quit, goal reached
11. Move to m along obstacle boundary
12. **IF** (obstacle is encountered at m in direction of g)
    THEN quit // goal not reachable
13. **ENDWHILE**

Figure 9.4: Bug 1 Algorithm (adapted from Ng, 2010)

The Bug 1 algorithm searches each encountered obstacle for the point that is closest to the target. Once that point is determined, the robot evaluates whether it can drive toward the target or not. If it cannot, the target is unreachable. If it can, the robot knows that, by leaving at that point, it will never re-encounter the obstacle.

The Bug 1 algorithm is sound; i.e., the algorithm will always find a path to the goal. The algorithm is also complete; i.e., the algorithm will find all possible paths.

9.4 BUG 2 STRATEGY

The Bug 2 basic strategy is to follow a line called the m-line, which is the line that connects the start point and the goal point.

The robot moves toward the goal along the m-line. If an obstacle is encountered, it moves along the perimeter of the obstacle until it reaches the m-line. Once it identifies the m-line, it resumes from that point to reach the goal. This process is repeated for all the obstacles until the goal is reached. Figure 9.5 is an example, and Figure 9.6 outlines the algorithm.

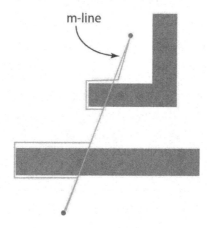

Figure 9.5: An example illustrating the Bug 2 algorithm (adapted from Choset et al. 2005)

```
 1. LET L = line from r to g
 2. WHILE (TRUE)
 3. r = robot's current location; g is the goal location
 4. REPEAT
 5. Move from r toward g
 6. UNTIL ((r == g) OR (obstacle is encountered))
 7. IF (r == g) THEN quit // goal reached
 8. LET p = r // a contact location with obstacle
 9. REPEAT
10. Follow obstacle boundary
11. r == robot's current location
12. m = intersection point of r and L closer to g than distance (p,g)
13. UNTIL (((m is not null) AND (distance(m,g) < distance(p,g)) OR (r == g) OR (r == p))
14. IF (r == g) THEN quit // goal reached
15. IF (r == p) THEN quit // goal not reachable
16. ENDWHILE
```

Figure 9.6: The Bug 2 Algorithm (adapted from Ng, 2010)

Bug 2 is less conservative than Bug 1 because the robot can leave earlier, due to following the m-line. A modification of Bug 2 is available (Horiuchi and Noborio, 2001). Antich and Ortiz suggested a variation they called Bug 2+ (Antich and Ortiz, 2009).

Typically, Bug 2 reaches the goal using a shorter path than Bug 1. Therefore, Bug 2 outperforms Bug 1. However, in some environments, Bug 1 outperforms Bug 2. For example, Figure 9.7 shows a case in which Bug 1 outperforms Bug 2, assuming the rectangle perimeter is twice the crescent perimeter. Bug 2 will travel along almost three sides of the rectangle, and not much of the perimeter of the crescent, plus the length of the m-line. However, Bug 1 will only travel around the crescent 1.5 times.

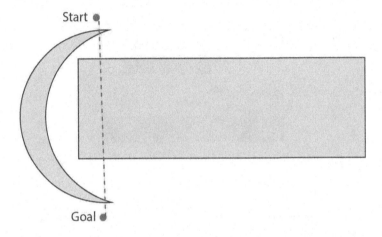

Figure 9.7: An example, in which Bug 1 beats Bug 2 for a left-turning robot

9.5   TANGENT BUG STRATEGY

The Tangent Bug algorithm was developed by Kamon, Rivlin and Rimon (Kamon et al., 1998). It uses distance sensors to build a graph of the robot's immediate surroundings and uses this to minimize path length. Figure 9.8 shows distance sensing. Consider the robot with a 360° range sensor, which can determine distances to all obstacles around it. The distance sensor emits probing rays, and, when they hit the object, they can form a tangent line to the boundary of the object. The periphery points on the tangent line touching the object boundary are called *discontinuity points* (DP). When a robot is using the tangent bug, it moves toward the goal. If an obstacle is encountered directly between it and the goal, it finds a continuous interval to travel along toward the end that will minimize the remaining travel to the goal. It compares points to DP on the way to the goal, and selects the shortest distance.

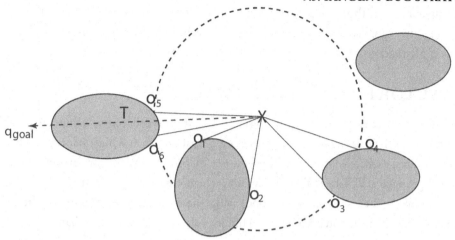

Figure 9.8: Robot location is marked by X. The dotted circle is the range of its proximity sensor (adapted from Choset et al., 2005)

In Figure 9.9, the robot continually compares d1 + d2 with d3 + d4, and decides which path to take. Tangent Bug uses sensors heuristics to compute the distance between the robot to DP and DP to the goal.

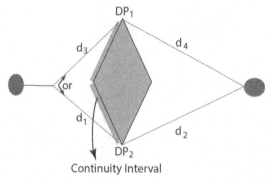

Figure 9.9: An illustration of discontinuity points

9.6   SEARCH NAVIGATION

In this type of scenario, we assume that the goal will be equipped with a sensor that continuously shows its presence. In a similar way, the robot will be equipped with a sensor that receives a signal from the goal, so that it can identify its presence. Furthermore, the sensor should give an intensity (i.e., its closeness) signal; otherwise, the robot will be trapped among the obstacles. With the goal sensor available, a bug-like algorithm can be used for navigation.

The major flaw with this approach is that the robot may be trapped in a situation like a box, where it finds the signal strength to be very high. A technique for navigating around obstacles is Vector Field Histograms.

9.7 VECTOR FIELD HISTOGRAMS

Consider a 2D Grid (a map) as shown in Figure 9.10. At each cell, sensor values are combined to encode certainty of the existing object. Low values are least certain and high values are more certain. Using these values, each cell location will produce a PUSH-AWAY vector for the robot, whereas the goal will produce a PULL-TOWARD vector, which creates a pull for the robot to reach the goal. By combining all these vectors, the robot will generate a steering histogram. The effects are very similar to the artificial potential fields techniques already discussed in Chapter 7.

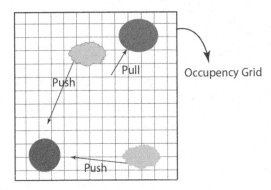

Figure 9.10: An occupancy grid

By using the steering histograms, the robot will be steered toward the least obstacles zone (valleys), so as to reach the goal. These obstacle valleys can be classified in two ways. The wide and narrow valleys, as shown in Figure 9.11, can be determined by using the threshold that needs to be adjusted experimentally.

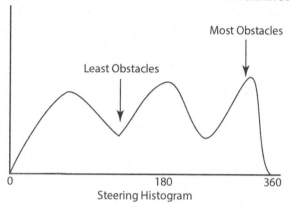

Figure 9.11: A steering histogram

9.8 SEARCH AND COVERAGE

A *coverage* algorithm produces a path that a robot must travel on in order to "cover" (i.e., account for by traveling over) the entire surface of its environment. Applications of coverage are ubiquitous, including vacuuming and sweeping a room, painting an area, searching for a lost item, security patrolling an area, and verification of a map's accuracy. Completeness of coverage can be ambiguous, even if complete coverage is guaranteed (González et al., 2005). The simplest approach is to simply travel in some fixed direction (e.g., North) until an obstacle is encountered, then turn around and continue coverage in pieces leading to coverage in strips. This approach can be problematic due to odd obstacle shapes and border angles.

There might always be some errors in terms of coverage. Coverage will miss areas close to edges and in corners, which can be remedied by allowing overlapping coverage of strips. Dividing the environment into smaller pieces will also help. For most applications being close enough to the obstacles is sufficient; i.e., sometimes, a rough coverage is good enough. It is possible to direct sensors to detect obstacles from a certain distance away without contact. Cell decomposition techniques in Chapter 8 are useful in dividing the areas into smaller segments for coverage (Choset et al., 2005).

Consider an environment for the purpose of searching for other robots, fire, intruders, or identifiable objects. Search robots may have sensors with various ranges and capabilities. Consider a simple environment with no obstacles, and a robot with omni-directional sensing with unlimited range capabilities. Such a robot can search without moving. The *kernel* of a star-shaped polygon shown in Figure 9.12 is the area of the polygon from which the robot can scan the entire boundary of the environment. If the environment is not star-shaped or has obstacles, then the kernel is empty (i.e., can't scan the entire environment from one location). We then need to determine a set of locations (scan points) that cover the entire environment. Placing a robot at each reflex vertex will ensure complete visibility coverage. Finding fewer observation points is a problem that is called the *guard placement problem* or the *art gallery problem* (Aggarwal, 1984). For a simple polygon environment with n vertices, locations are occasionally necessary and always sufficient to have every point in the polygon visible from at least one of the locations.

Figure 9.12: A star shaped polygon

In order to search for all $\left\lfloor \frac{n}{3} \right\rfloor$ things, the robot must navigate extensively in the environment. In keeping with efficiency, the robot may seek the shortest possible path in the environment such that the robot covers (senses) all areas in the environment. In computational geometry, this has been termed the *shortest watchman route* (Carlsson et al., 1993). This is a difficult problem. Therefore, approximations are acceptable. One method for addressing this problem is to find guard placement locations and then connect them with an efficient path; i.e., the robot will travel between multiple goal locations. Given a number of locations that the robot must visit, the cheapest round-trip route that visits each location once and then returns to the starting location is called the *traveling salesman problem* (Applegate et al., 2007). An approximate tour is based on a minimum spanning tree from the start location, with a running time of $O(n^2)$ for n locations.

9.9   CONCLUSIONS

Since the environment often changes over time, a map is needed to represent the environment and relative locations of the robot, obstacles, and goals. Maps are often estimates that guide the robot. Sensors are needed for local navigation and accuracy.

There are several map types. Topological maps maintain relationships between obstacles and free space in the environment. Obstacle maps take the obstacles vs. accessible locations. Free-space maps track locations for safe navigation that is the opposite of the Obstacles. Path maps are often used in the industry to track safe routes in the environment.

CHAPTER 10

Multi-Robot Mapping: Brick and Mortar Strategy

The inspiration for this strategy is that there is a need to explore an area in a minimum amount of time, in emergency situations where it's too dangerous for humans. Japan's 2011 nuclear incident is a perfect motivating example. In this chapter, we will describe a strategy called *brick and mortar* for multiple robots collaboratively constructing a map (Ferranti et al., 2007).

The area is divided into square cells, with cells that can be in one of the following four states:

1. **Wall:** cell can't be traversed due to the presence of actual brick or other obstacle

2. **Unexplored:** no one has visited the cell

3. **Explored:** the cell is traversed at least once, but it might need to be used to reach other cells

4. **Visited:** the cell is traversed and is not useful for reaching unexplored areas

A few setup assumptions are essential. Initially, agents are deployed on boundary cells as if entering the area. At each step, they can go to a north, south, east, or west neighboring cell. Agents don't know their location with respect to the environment (e.g. there is no GPS available). They cannot use wireless communication to co-locate each other. Agents can communicate indirectly by marking information in cells. This style of communication is called *Stigmergy*. No central control points exist. Typically, cell information is stored locally in Radio Frequency ID's (RFID). Agents use local information for independent decisions.

10.1 ANTS ALGORITHM

At the start, all cells are marked with a zero (i.e., marked as unexplored). At each step, an agent reads the values of its four neighboring cells and picks the least-visited cell to move into. Before moving, it updates the current cell (i.e. increments visited value by one). All cells are eventually visited. This strategy has a few advantages: it is simple, no map is needed, and if an agent is relocated, it can continue unhindered; and it is flexible and fault tolerant. Ants approach also has shortcomings: since cells are not marked explored (i.e., to signify completion), there will be redundant visits; there is no way to determine completion—agents work until they are out of energy; agents are inefficient and there are redundant explorations; collaboration levels are low; and many explore old cells, while only a few go to new ones. There are strategies to remedy problems and the following section will discuss these strategies (Ferranti and Trigoni, 2011).

10.2   MULTIPLE DEPTH FIRST SEARCH (MDFS)

With one agent, the robot builds an exploration tree. When moving to a new cell, it records the parent cell and direction of movement (N, S, E, W) to reach that cell. When there are no unexplored neighbor cells, the agent is at the end of a branch (i.e., a leaf node). It will mark the current cell visited and return to the parent cell, looking for an unexplored neighbor cell. In this way, all cells will be marked at least once. Going down the branch, cells are marked as explored. Going up the branch, cells are marked as visited. If the agent reaches the root node and all cells are marked visited, the algorithm terminates. With multiple agents, each agent builds its own DFS tree, but also places its own ID in the cell. Agents will stay in their own trees, unless they exhaust unexplored cells. Then they will enter trees of other agents, marking them accordingly (see Algorithm MDFS in Figure 10.1).

1. **IF** the current cell is unexplored, **THEN** mark it as explored
2. Annotate the cell with your ID
3. Annotate the cell with the direction of the previous cell (parent cell)
4. **END IF**
5. **IF** there are unexplored cell around, then go to one of them randomly
6. **ELSE IF** the current cell is marked with your ID, **THEN** mark it as visited
7. Go to the parent cell
8. **ELSE** go to one of the explored cells randomly
9. **END IF**
10. **END IF**

Figure 10.1: MDFS algorithm (adapted from Ferranti et al., 2007)

Advantages for MDFS are that it provides faster speed with a lower redundancy rate. However, MDFS suffers from several disadvantages: there are possible overlapping explorations, it is not efficient in open areas, and an agent might become trapped before the search is completed.

Multi-robot coordination is not limited to map generation. Unmanned aerial vehicles (UAVs) are used in a variety of tasks including intelligence and search and reconnaissance (ISR) (Shen et al., 2008). In addition, a target detection algorithm is available that determines whether or not a target object exists in a video frame (Symington et al., 2010).

10.3 CONCLUSIONS

The brick and mortar algorithm has introduced a novel class of collaborative mapping. It exemplifies *stigmergy*, which is indirect communication for coordination. This is a seminal work that can be the basis for a promising multi-robotic ISR in development.

PART III

State of the Art

CHAPTER 11

Multi-Robotics Phenomena

Chapter 10 illustrated mapping with a team of identical robots. Chapter 11 will go beyond mapping and discuss capabilities and limitations encountered with multiple robots. For a number of years, the U.S. Department of Defense promoted research about creating robots that behaved in formation (Balch and Arkin, 1998). More recently, heterogeneous robots were considered as teams. Balch and Parker (2002) suggested correlating heterogeneity with performance. Hierarchical *social entropy* was developed as an application of Shannon's information entropy metric to robotic groups that provides a continuous, quantitative measure of robot team diversity (Shannon and Weaver, 1971). The metric captures important components of the meaning of diversity, including the number and size of behavioral groups in a society, and the extent to which agents differ. The utility of the metrics is demonstrated in the experimental evaluation of multi-robot soccer and multi-robot foraging teams.

An extension of social entropy is a metric that differentiates a group with agents at different power levels. Tucker Balch's social entropy differentiates the agents with different group sizes (Balch, 1998). Let R be a set of robots. C is a classification of R into c, possibly overlapping subsets. Ci is an individual sub-set of C as in $\{R_1, R_2\}$. *Heterogeneity* of a group of robots (i.e., social entropy) is denoted by H(R).

$$H(R) = - \sum_{i=1}^{c} p_i \times \log_2(p_i) \quad (1)$$

$$\text{where} \quad p_i = \frac{|c_i|}{\sum_{j=1}^{c} |c_j|}.$$

Using H(R) metric, H(R) = 0 if robots in R are all homogenous (i.e., functionally equivalent). If robots in R are all uniquely different from one another, H(R) has the maximum value.

The principle of social entropy (i.e., heterogeneity) is extended to differentiate among groups having different power profiles. For simplicity, the *power homogeneity* is represented as PH, and is computed by the following expression (Hexmoor et al., 2008). Here, m is the number of groups. p_i is defined as the ratio of the number of agents in a group i over the total number of agents. w_i is defined as the ratio of power level of the group over the maximum power level of all the groups.

$$PH(R) = \sum_{i=1}^{m} (P_i + w_i) \times \log_2(P_i + w_i) \quad (2)$$

Advantages for multiple robots include capacity for a larger range of tasks, greater efficiency, improved system performance, fault tolerance, lower economic costs, ease of development, and distribution of sensing and acting. Distributed artificial intelligence and multi-agent systems are allied disciplines that consider how tasks can be divided among robots that share knowledge about the evolving problem space (Wooldridge, 2009). Biological inspirations have inspired the development of robots that follow simple reactive rules, and the interaction between robots produces unforeseen, emergent behaviors (Bar-Cohn and Brazeal, 2003).

An important capability for multirobotics is learning without human supervision in order to solve problems. They must be able to deal with dynamic changes in the environment, as well as changes in their own performance capabilities. Robots might need to learn and to evolve in order to increase the likelihood of survival and improved task performance. Adaptation is how a robot learns by making adjustments, whereas learning is behavioral change to adapt to the environment.

Evolution is the process of selective reproduction and substitution based on the existence of a distributed population of vehicles. Evolution may not perform well when certain environmental changes occur that are different from evolved solutions. Learning is a set of modifications taking place within each individual during its own lifetime. Learning often takes place during an initial phase, when task performance is considered less important. The control policy is used to produce reasonable performance.

Foraging is a common problem studied in robotic colonies. Foraging is collecting items and possibly bringing them to some specific location and there are many variations of this problem. Foraging performance can be improved by reducing collisions or interference among robots, preventing robots from traveling over the same areas, and directing robots toward collections of forage items.

Communication is another important issue for multirobot algorithms. Consider robots that might be structured in a hierarchy where each robot belongs to a group and all group members can communicate to a group leader via wireless communication. The leaders are grouped together with a higher-level leader to which they communicate.

Within a hierarchy, worker robots must always remain within communication range and allow data to be transmitted to the leader (e.g., map data). They must allow the leader to send commands at any time (e.g., new directions and updated task assignments), and they must be allowed to dock rapidly for energy replenishment. This is called *hierarchical communication* (McLaughlan and Hexmoor, 2009).

Lower-level robots can be given global knowledge of the environment and/or of task completion. Hierarchical communication provides benefits over no-communication schemes for more complex problems. It can allow the steering of robots to accomplish tasks more efficiently.

Another investigation that overlaps multirobotics is swarm intelligence and morphogenesis. These concepts will be discussed in the following section.

11.1 SWARMS

Swarm intelligence explores the collective behavior of decentralized and self-organized systems (Eberhart and Kennedy, 2001).

Emergent behavior is when a number of simple entities (agents or robots), operating in an environment, form more complex behaviors as a collective. A great deal of inspiration is derived from observations of social organisms such as ants. Ant colonies found in nature send forager ants to search for food. As each ant searches about the environment, it leaves behind samples of *pheromone*. When it finds a food source it takes a piece back to the nest, reinforcing the pheromone path it took to get there. The majority of all ant species are blind, and as a rule will follow the strongest pheromone trail to direct them between the food source and the nest.

Ant simulation algorithms are used to find optimal solutions by moving through a space representing all possible solutions. Instead of using pheromones, simulated ants simply record their positions and the quality of the solution, determined by time or steps it took to reach the solution (Dorigo, 2004). Applications of this algorithm are in vehicle routing, the traveling salesman problem, and genetic algorithms.

When food sources are large, as shown in Figure 11.1, it is sometimes easier to carry a piece of food to the entrance of the nest as a group. This way the whole swarm can dismantle it more safely.

Figure 11.1: Emergence of cooperation among ants

Morphogenesis is the process of autonomous mobile robots self-assembling into specific morphologies in order to adapt to their environment and perform a task. Robots assemble into a particular morphology as a response to rules that define the intended morphological shape. When obstacles are encountered, the discovering robot initiates communication to form a structure. When one robot connects to another, the newly connected robot receives instructions on how it will continue the local structure. In turn, it will then attract other robots for connection. The process is repeated until the structure is complete. Claytronic morphing robots are microscopic swarms of robots capable of morphing into any shape (Kirby et al., 2007).

11.2 CONCLUSIONS

Using heterogeneous groups of robots may prove to be a robust and efficient solution for large problems. Experimentation will help to adjust the use of multiple robots for desired solutions. The initial notions of social entropy and power homogeneity, for characterizing the social nature of robots, are part of the social informatics movement (Kling et al., 2005).

CHAPTER 12

Human-Robot Interaction

Humanoid in form or not, human interaction with robots (HRI) has long implied mighty, but subservient, robots in the popular culture. This is depicted in the K9 role in BBC's science fiction series *Dr. Who* (shown in Figure 12.1a); the robot from the U.S. television series *Lost in Space* (Figure 12.1b); and the U.S. Navy's SPAWAR robot prototype named Robart, an armed robot that can track anything that moves (Figure 12.1c). An interesting and illuminating early example of human-robot interaction has been the robotic wheelchair system (Yanco, 2000)—here, interactions demonstrate negotiation over degrees of autonomy that the human has granted to the robot (Yanco and Drury, 2004).

These perceptions are rapidly shifting and blur the distinction between humans and robots. In one dimension, robots are perceived as indefatigable worker bees. In another dimension, they have legal rights and responsibilities, just as human citizens. The Japanese space agency, JAXA, has plans for a permanent base on the Moon by 2020, for robots, built by robots. A variation of this idea was echoed by the 2012 U.S. presidential candidate Newt Gingrich, who suggested creation of a lunar colony that he said could become a U.S. state. Started in 2006, HRI as a nascent discipline is promoted by the conference series sponsored by major professional organizations such as the Association for Computing Machinery (ACM) and the Institute of Electrical and Electronics Engineers (IEEE). Given the latest visions and trends in robotics developments, HRI must address the fundamentals of interaction between robots and humans. Establishing equivalency is inevitable in the future. There is no doubt that technology can and will support peer-level interactions.

a.   b.   c.

Figure 12.1: Subservient robots from science fiction popular culture. (a) "K9" robotic dog from BBC series *Dr. Who*, Reproduced under license from BBC News / BBC Sport / bbc.co.uk - © BBC. (b) "Robot" from U.S.television series Lost in Space, (c) The U.S.Navy's SPAWAR robot prototype named Robart. Copyright © 2008, Dave Bullock. Used with permission

Technological advances will be twofold. On the one hand, robots will gain more sophisticated and natural (perhaps anthropomorphic) demeanors. On the other hand, humans will be outfitted with physical and cognitive prosthetics to augment their capacities for perception and cognition. Evidence of the start of this movement is the rapid rise in human reliance on smart phones. Current state-of-the-art vision and hearing prostheses are producing emerging products that will be commonly used a decade from now. Augmenting senses of touch, taste, and smell are lacking in research attention. However, we forecast significant developments in these areas in the near future.

Current modalities of HRI include tactile and joystick control, as in the use of the teach pendant for programming manipulator trajectories (See Figure 6.9 in Chapter 6). Another example of a hand-held remote control is shown in Figure 12.2.

Figure 12.2: The VEX robot kit's VEXnet Wireless Joystick. Copyright © 2003-2013, RobotShop Distribution inc., http://www.robotshop.com. Used with permission

Figure 12.3: Romo, an Android smart phone affixed to a moving robot platform. Retrieved from http://www.kickstarter.com/projects/peterseid/romo-the-smartphone-robot. Copyright © 2011, Romotive, http://romotive.com/. Used with permission

Audio control is another common method for contemporary HRI with Romo the robot (see Figure 12.3). Using speech recognition software on a smart phone, robot control surpasses remote-control devices. We predict exciting developments with speech control. In the next section we will outline multimodal interactions and end the chapter with techniques for pointing.

12.1 MULTIMODAL INTERACTIONS

In field robotics, often there is a need for a human operator, to provide high-level instructions simultaneously to teams composed of semi-autonomous robots and humans. Communicating with robots is not practical with input devices, such as a keyboard or mouse. Touch-based input is becoming more prevalent (Hayes et al., 2010). Based on NASA experiments, multi-touch interface is shown to lower overall workload and reduce interaction frustrations (Humphrey and Adams, 2011).

A fascinating activity is interaction between apes and robots. Figure 12.4 shows a robot prototype called *robobonobo*. Sanctuary resident apes are trained to use an Apple iPad application to control this robot, which will have a water gun on board. People are also allowed to communicate with the apes using the application.

Figure 12.4: Robobonobo (image courtesy of Dr. Kenneth Schweller, chair of the Great Ape Trust Board at Great Ape trust, Des Moines, Iowa)

Recall applications of *affect* from Chapter 4. Models of human emotions provide reasonable and useful heuristics for robots as rational agents. It is argued that incorporating emotions will reduce and control non-determinism. It facilitates flexible cooperation and coordination among robots and it also provides efficacious human-robot interfaces (Steunebrink et al., 2007). Figure 12.5 shows the robot iCAT reacting to a person with emotional responses.

Figure 12.5: The iCat robot, developed by Philips, is an Experimentation platform for human-robot interaction (image courtesy of Philips Research)

12.2 POINTING

Pointing-based human-robot interface for robot guidance is a rather novel research front (Cipolla and Hollinghurst, 1996). Another effort offered an algorithm for disambiguating pointer referents (Hexmoor and Yang, 2000). The algorithm operates in three phases. Six objects and one pointer are shown in Figure 12.6. In the first phase, either the pointer tip clearly touches the object, or the pointer cannot possibly select certain objects. Objects 1 and 2 are eliminated, since they are outside the pointer's fields of regard. In fact, any object that is located behind the plane at the pointer tip, and perpendicular to the pointer line, is not considered. An exception is when the pointer tip is literally touching or within infinitesimal proximity to the object. In that case, the object is selected and we are finished.

In the second phase, we consider the region in the range and let's assume there is no field of regard. Figure 12.6 shows objects 3, 4, 5 and 6 in this region. In this phase, we select the object that forms the smallest angle to the pointer line. If there are objects with the same angle to the pointer line, the object that has the shortest distance to the pointer tip is selected. Naturally, object 6 is the most preferred, subsequently object 5 is selected. If object 5 and 6 did not exist, we would have to choose between objects 3 and 4. We would choose object 4, since it has the smallest angle.

In the third phase, we consider the field of regard in the range. We assume vision cannot select objects based solely on angles. In this range, perpendicular lines (PL) from objects to the extended pointer line (i.e., indefinitely elongates pointer vector) are considered. Next, consider the ratio shown in the following equation that we call candidacy metric for each object O.

$$\text{candidacy } (O) = (PL(O) - \min(PL))/PL(O) \quad (1)$$

PL(O) is the length of the PL segment for the object O. min(PL) is the PL for the object that produces the smallest PL length. For each object O, candidacy (O) is compared to the ratio of angles $|\alpha - \beta|/\alpha$, and, if it is smaller or equal, we say the object is a candidate. After considering all objects, if there is a single candidate, the object is selected; otherwise, the situation is ambiguous and no object is selected. This strategy has shown preliminary promise but further refinements are needed (Hexmoor and Yang, 2000).

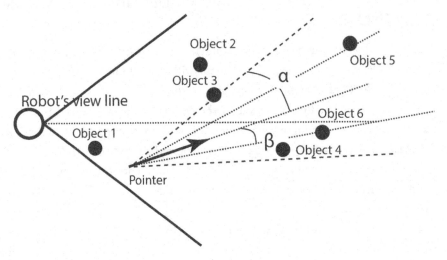

Figure 12.6: A simplified 2D diagram depicting top-down view of a pointer and several objects (Hexmoor and Yang, 2000)

12.3 OBSERVATIONS AND CONCLUSIONS

HRI emerged as a discipline with the inception of the ACM/IEEE annual conference series in 2006. With a shift in the perception of robots as artificial human counterparts and the narrowing of the gap between human and robot sensorimotor capabilities, early HRI issues are coalescing into issues of cohesion and synergy; i.e., collaboration and coordination among teams of peers (Dias et al., 2007). Mind-machine interface has been a slow and steady research discipline since the 1970s, largely targeting control of prosthetic limbs. Basic information can be found at the IEEE Spectrum magazine: htttp://spectrum.ieee.org/brainmachine0312. We must stay tuned for future developments.

CHAPTER 13

Fuzzy Control

by Henry Hexmoor and Anthony Kulis

Lotfi Zadeh introduced *fuzzy logic* in 1965 (Zadeh, 1965). Fuzzy logic is a representational framework that captures imprecision of expression in human natural language. We begin with the rudiments of fuzzy logic and then turn to applications for robot control. Fuzzy logic is based on the idea that symbols, or linguistic variables, have meanings that diverge from one entity to another. When a reader sees a person described as "average height" in their favorite book, they immediately create an image that is only relevant to them. Average to one reader may be 1.8 meters in height, but another reader whose stature is a towering 2 meters would immediately define 1.8 meters as short. In fact, the word "towering" is a linguistic variable that can be defined in fuzzy logic.

It is quite simple to see that these linguistic variables can overlap. It is this overlapping of meanings that constitutes the term *fuzzy*. If we use the height example, we also will invariably encounter a height that no one could possible say is average. We know that children approximately 1 meter tall would consider everyone tall, yet the average height, male or female, is around 1.7 meters. As we trend to the taller side of the height spectrum, fewer people encounter individuals taller than themselves, so the meaning of average has lowered importance. Figure 13.1 demonstrates a linguistic variable used in fuzzy logic. *Membership function* maps a magnitude quantifying a concept to the degree to which that value reflects saliency of the concept. As we can see from Figure 13.1, the greatest membership for tallness of a person is reflected when an individual's height is around 1.77 meters, but this is not a complete statement. We cannot only classify a person as being average in height when common sense also asks for short and tall. If we introduce these two new variables, a better picture of classification emerges.

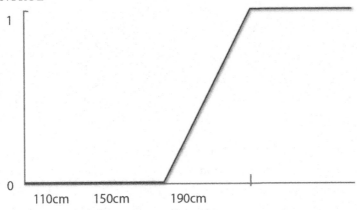

Figure 13.1: Example of a fuzzy membership for the concept of "tall"

Figure 13.2 illustrates a membership function when we also include short and tall people. For this example, we will consider ourselves experts in people height. We use expert knowledge to say that if a person is 1.77 meters tall, they are average in height. But as we deviate from 1.77 meters toward 1.98 meters, fewer people will suggest average to be a good descriptor of height.

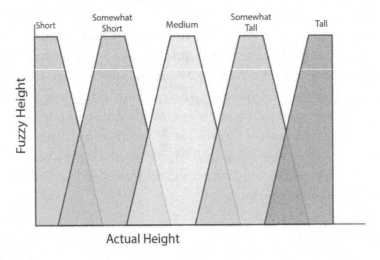

Figure 13.2: Another example of membership functions: people consider overlapping height average concepts rather than a single function

Instead, it is more realistic to consider overlapping set of functions representing concepts that are imprecise. This is shown in Figure 13.2. It is this overlapping that gives fuzzy logic its strength. This overlapping can formally be described as *noise*, and noise can influence a discrete system to make incorrect calculations. While we say 1.77 meters is average, it also is fairly logical to say 1.78 meters is average, but, as in Figure 13.2, when a person's height is too different from our expert analysis of an average height, it becomes too difficult to discretely determine, so a *fuzzy membership value* is derived. If we also have information about a person's weight, we can then use these two pieces of information to infer something about his/her physical fitness. The technique of modeling we choose to demonstrate is called *Mamdani Fuzzy Inference*. *Inference* in fuzzy architectures is a common approach, and can be found in literature by Safiotti (1997), Lakov (2002), and Chakraborty (2008).

Locomotion in robots was reviewed in Chapter 1, covering legs that step across their environment, and wheels that roll across. In an ideal environment, a robot can use discrete methods to control their appendages for movement. If a robot's task is to merely follow a straight line on the floor, developing a steering control would be a relatively easy task. If a 90° turn is introduced, the algorithm to control steering would become more complex. As we continue to introduce other turns of varying degrees, or even odd-shaped obstacles, the amount of work an algorithm requires grows dramatically. What increases is not the complexity of making the appropriate navigation technique, but rather the techniques for handling the noise in the system. We may introduce another dimension of complexity when we realize that our sensors are only partially accurate, and that sensor errors can reduce discrete navigation algorithms to novelty concepts rather than real-world capable ones.

Saffiotti (1997) provides the details of three models of robot locomotion control. The base model does not use fuzzy control, but provides a framework in which fuzzy control can be introduced. Figure 13.3 illustrates this model. The hierarchical model has three phases. First, the robot must take a snapshot of its environment through a sensor. The information captured by the sensor(s) must be modeled into an information vector that can be acted upon. As the environment image is formed, a plan must be constructed. Once planned, the robot must execute the plan.

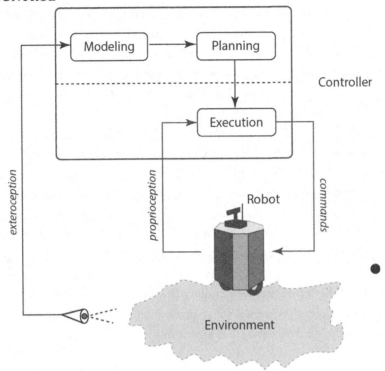

Figure 13.3: Saffiotti Fuzzy Control Model, (adapted from Saffiotti, 1997)

The planning method has been successfully used in numerous applications, but is quickly being outdated due to the dynamic nature of the world and the limitations of this model to laboratory work (Saffiotti, 1997). We can clearly see that, if a sensor captures a snapshot of the world, but, if while in the modeling phase or the planning stage, the world changes, the execution stage is out of date and would need to be refreshed. Saffiotti accounts for real-time sensor data only at the execution stage. However, new information about the world requires returning to the modeling phase, or could otherwise result in conflicting sensor data during the execution stage. By introducing fuzzy membership functions, changes in the environment can influence the execution phase without the need to remodel (Saffiotti, 1997). Adding this new loop to the hierarchical model increases the complexity of the execution stage. To handle the complexity, many divide this component into smaller subsections known as *behaviors*. Through fusion of data, behaviors, such as following a line or avoiding an obstacle, can be handled flexibly. Many strategies exist for fusion of data from subsumption architecture to fuzzy control.

Proprioceptive behavior is a common fuzzy strategy where the robot forms a control sequence to navigate a path as closely as possible. Fuzzy control in potential fields provides added capabilities. Each one of these techniques found in proprioceptive behaviors provides components, which, when used collectively, can provide robust and comprehensive robot locomotion control. Similar to proprioceptive behaviors, Saffiotti also offers *sensor-based* behaviors.

Hybridization of techniques allows us to extract beneficial components of one subset of artificial intelligence methods to replace unfavorable components of another. The following section of our fuzzy logic overview introduces some techniques to form complex behaviors in fuzzy robotic control.[1]

Tunstel and colleagues (1996) introduced two models of fuzzy systems using genetics. Their approach was to provide a system using GPs for the purpose of navigation and a system using GAs for manipulation of controls. In the first of their two systems, a complex behavior scheme is managed by creating a parse tree of inference rules that provide the policy for navigation through an environment. While the membership values do not change, the rule base is genetically evolved to find an optimum tree that represents the if-then rules found in Mamdani Fuzzy Inference. The fitness of the rules that are chosen is determined by a Euclidean distance measure of error, so that those inference rules with poor navigation are weeded out while the rules with optimal results are bred into future generations (Tunstel et al., 1996). When evolved offline, the rule base that is genetically superior is used online, and demonstrates quick error correction to return to the proper heading when the robot is started from arbitrary positions. The approach described by Tunstel et al. does not evolve complex control behaviors, but rather illustrates how the fuzzy rule set can be evolved without an exhaustive approach. When we introduce complex policy behaviors, this approach provides a framework to reduce offline development where exhaustive approaches are impracticable. While GPs provide a way to evolve rules, GAs provide a way to evolve the values of linguistic variables.

The second approach found in Tunstel et al. (1996) describes uses of GAs to develop variable values for environments where expert knowledge cannot be accurate. We could imagine how there is not an expert on Earth that truly understands the navigation environment on Mars. By using GAs online to learn what is small or large for the conditions found on Mars, the fuzzy rules can be affected because their values change. It can be assumed that a large rock on Mars is a navigation hazard the same way as it is on Earth, but the meaning of large on Earth does not have to have the same meaning as it does on Mars.

This work is further detailed in Homaifar et al. (1999) and describes the learning requirements found in their navigation system. By using fitness functions similar to GPs, GAs can evolve online, providing a temporal reasoning to robot control. When realizing that the robot sent to Mars will probably never come back to Earth to be reconfigured, and that there may be times when the robot is out of radio contact with operators on Earth, using online GAs to modify fuzzy variable values makes common sense. In the example provided by Tunstel et al., GAs were used to teach the fuzzy membership values regarding the meaning of error and overshoot when trying to navigate a

---

[1] Fuzzy Systems using Genetics: Genetic AI is commonly referred to as Genetic Algorithms (GA) and Genetic Programming (GP). For a review of these topics, see Floreano and Mattiussi, 2009.

course. They found that using GAs in fuzzy control improved correction time by 11% over non-GA fuzzy controllers (Tunstel et al., 1996). In this sense, GAs were used to fine-tune fuzzy controllers.

We have robots in every part of our lives, but their image in our minds remains as machines and not entities with significant importance. As humans communicate, we not only have the formal constructs of language, but also the ability to sense meaning from these constructs. Kim et al. (2003) report on a human-robot interaction system for the disabled, to assist in everyday living, that tries to bring meaning to formal constructs of interaction. Currently there are two major systems of interaction they have developed. The first, a robot arm that assists in drinking, uses fuzzy logic to decide whether the human has opted out of the drinking interaction after the initiation of the robot's action (Kim et al., 2003). As the robot lifts the drinking cup toward the human's mouth, an intelligent visual sensor system observes the human's body language. Kim et al. decided that a continuously open mouth as the robot's arm approaches is a behavioral reinforcement signal to continue the robot's action. If the human's mouth starts to close, this behavior cancels the motion. As we have already demonstrated in previous subsections, it is difficult for a discrete system to decide how much opening is open and how much closing is closed. Using fuzzy logic, Kim et al. were able to achieve a 92.2% success ratio when determining intention within this problem domain.

Another fuzzy interaction Kim et al. employed was to recognize emergency situations. Because the robot was interacting with the human in a physical sense, emergency reactions were needed to prevent the robot from accidentally injuring the human. To capture the emergency input, vital signals were used as inputs because it is assumed that the human would have a muscular reaction to the presence of danger. When the average signal of the value was found to be at dangerous levels using a fuzzy classifier, the robot was able to estimate the level of emergency and perform the appropriate safety action. The success rate using three fully-abled people and one quadriplegic was 100%.

13.1 CONCLUSIONS

Fuzzy logic has provided representational capabilities to model robotic operations in the language of a human expert describing those operations. Thus far, this has been demonstrated in navigation and a few common human-robot interaction tasks. From tedious and delicate surgical procedures to large movements in construction, it is conceivable to begin with a description of the operation by a human expert and develop counterparts of it in fuzzy models suitable for a robotic mimicry.

CHAPTER 14

Decision Theory and Game Theory

Decision theory (DT) is used by self-interested persons to make optimal decisions in uncertain environments (Raiffa, 1968). DT is often likened to a game against nature, in which choices of nature are considered random. When the opponents are other self-interested, independent individuals, DT becomes a multi-person DT. This is the premise for *game theory* (GT). For a majority of GT, all individuals are strictly interested in their own welfare and there is competition among them to maximize the benefits from their choices. The methodologies of game theory provide a language to formulate structure, analysis, and understanding strategic scenarios.

GT has been applied to problems of multiple robots coordination, high-level strategy planning, and information gathering through manipulation and/or sensor planning, and pursuit-evasion scenarios (Turocy and Von Stengel, 2002). A GT formulation of robotic tasks is available from (Lavalle and Hutchinson, 1993).

A multi-robot searching task for multiple targets has been modeled as a multiplayer, cooperative nonzero-sum game (Meng, 2008). Each robot solves a static game at each time step, producing an overall sequential solution, while the robot team continually searches for scattered targets. With N robots, the game matrix is an N dimensional matrix. Strategies are search region choices. In the simplest case of two robots, when they are both free or only one is free, search decisions must be made. To make the problem tractable, each robot creates a decision buffer to store the new decision whenever a new game starts. Region numbers are stored in the order of the searching sequence. Once a robot finishes its current region, the region number is removed from the buffer. When the robot is free, it will determine a new decision. When the robot is not free, it will continue with its current search. Figure 14.1 shows the flow chart explaining the conditions involved in this decision-based approach for when a game needs to be solved (Meng, 2008).

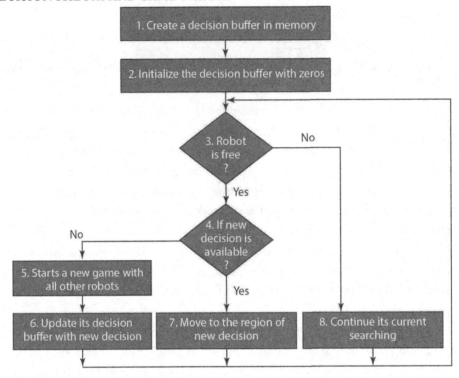

Figure 14.1: Flowchart of a decision-based approach (adapted from Meng, 2008)

The *Pursuit and Evasion* problem is important in robotics for finding an intruder in a given environment (Chung et al., 2011). The problem of pursuit is also referred to as tracking. The evasion problem can be considered as keeping multiple robots or obstacles from colliding, while executing tasks. In robotics, pursuit-evasion games are used for studying motion planning that arises in adversarial settings, such as catching intruders, burglars, and so on. In addition, pursuit-evasion games are used to obtain results on the worst-case performance of robotics systems. For example, imagine a patrolling setting in which the robots try to find an intruder. In this case, by treating the intruder as an adversarial entity trying to avoid being found, one can obtain worst-case bounds on the number of robots necessary for finding an intruder in a given environment. This is because pursuit strategy, if it exits, will guarantee that the intruder will be found (Chung et al., 2011).

In the remainder of this chapter we will refer to GT applied to security concerns among a multi-robot system in a dynamic environment. Examples of security concerns include protection of airports and other transportation systems, patrolling forests for protection of humans and wildlife from poachers and smugglers, and detecting the illegal flow of weapons, drugs, and money. Since there are often limited security resources, GT is used for adversarial reasoning for security resource allocation and scheduling problems.

In the application of robots to human security concerns, a robot would detect security concerns using sensors and actuators. Sensors allow it to monitor a portion of the environment, checking for the presence of an intruder. Actuators allow the agent to change the current view by focusing sensors on different portions of the environment. A patrolling strategy moves a robot in a given environment, in order to prevent intrusions.

Consider a particular range of environment that should be secured by mobile robots. The purpose of security is to protect the people and their properties from intruders, as well as to distinguish between intruders and non-intruders (Huang et al., 2008).

The main robot is called the *master*, and other robots are *slave robots*. The slave robots are placed in locations where there is a possibility of intruders entering. The position and range of the individual robot is assigned by the master robot and can be changed by it at any time, based on the circumstances. The master can communicate with all slave robots and slave robots cannot communicate with each other. Only the master holds the databases for identification and verification. This data is used to distinguish between an intruder and non-intruder. Slaves cannot make independent decisions until the master grants permission, or they'll be disabled. Consider a scenario where there is an unexpected attack. A slave robot can take action if necessary and sends an alarm signal to the master robot to take further action. This would also be the procedure in other emergency situations.

Another security application is the adaptation of Stackelberg games (i.e., the principal-agent games) as a specific class of GT games that are used to model patroller-intruder scenarios, in which both patroller and intruders have defined strategies. In application of the Stackelberg game, the patroller is allowed to observe the intruder's strategy before choosing its own strategy. Thus, there is an advantage for the patroller over the case where players must choose their moves simultaneously. This game allows pure strategy Nash equilibrium (Paruchuri et al., 2008).

Bayesian Stacklerberg games are Stackelberg games that account for probabilistic events (i.e., probabilities of encountering different types of intruders) and advocate randomized patrolling or randomized inspection strategies to counteract the probabilistic probabilities of intruder incursions. These algorithms are currently deployed in multiple applications and are leading to advances over previous approaches in security scheduling and allocation. For example, ARMOR (Assistant Randomized Monitoring over Routes) has been deployed at the Los Angeles International Airport to check points on the roadways entering the airport and canine patrol routes within the airport terminals (Paruchuri et al., 2008). ARMOR consists of algorithms for solving Bayesian Stackelberg games.

Although it is implemented in a number of scenarios, problems remain. The problem of finding an optimal strategy of patroller and intruder in the Bayesian Stackelberg games is NP-hard, and transforming the game into a normal form game using the Harsanyi style transformation loses the compactness of the game (Conitzer, 2011).

There are two types of agents (*game players*). The *leader* robot (the *patroller*) commits to a mixed strategy, in which strategies are selected with optimally determined probabilities. The *follower* robot (the *intruder*) is the second type of agent that responds optimally to the leader. The Bayesian Stackelberg game is a multiple follower-type game, which means a number of intruders (i.e., followers) may attack in a given environment.

The patroller's pure strategy is possible routes to be monitored and the strategy of the intruder is entry points to the environment. The patroller's mixed strategy is probability distribution over routes. By selecting routes probabilistically, the intruder will be unaware of a patroller's decision.

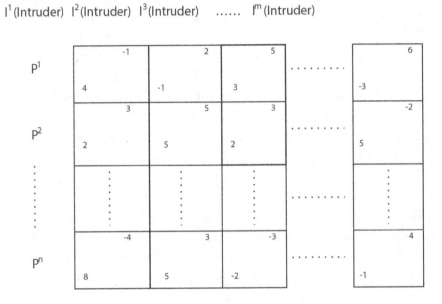

Figure 14.2: A depiction of n patrollers and m intruders

Strategies for n patrollers and m intruders are depicted in Figure 14.2, which represents the Bayesian Stackelberg Game between them. The values in the matrix are game utilities (i.e., rewards for combinations of strategy selection).

Consider patrollers p^1, p^2, p^3....p^n is given in the environment and target t^1, t^2, t^3 and t^n. Monte-Carlo sampling is used from the space of attacker types to estimate the probability that a target will be attacked for a given patroller strategy. Details of estimation are beyond our scope (Pitta et al., 2011).

14.1 CONCLUSIONS

In this chapter we presented brief references to game theory approaches for multi-robot search, pursuit-evasion, and security. Formulating problems in game theoretic approach provides a principled account of decisions robots must make and reasoning they are required to perform. Unfortunately, the robotic problems modeled with GT are often too large in structure and size to be computationally tractable. Large problems will often have multiple equilibriums (solutions) and choosing among them is a secondary complex problem. Adoption of GT in robotics has been a welcome tool and we hope for continued developments to help us address real-world problems.

PART IV
On the Horizon

CHAPTER 15

Applications: Macro and Micro Robots

Robots that are larger than an average human size and those smaller than an average human hand are in use in many applications, including space missions, military activities, and ocean projects. In this chapter, we outline a few promising prototypes. Since this is a fast-evolving domain, our coverage is not comprehensive and will remain incomplete.

Hippocrate and DERMAROB are a pair of robots designed by the LIRMM laboratory and manufactured by the SINTERS company in France. Hippocrate (Figure 15.1) is a robotic system developed to manipulate ultrasonic probes on the patient's skin for diagnosis in cardiovascular disease prevention. DERMAROB (not shown) is a robot dedicated to skin harvesting, which is a critical step during skin grafting surgery performed mainly on patients with burn injuries. DERMAROB holds the dermatome, a shaver-like device, which is the same as the tool used by the surgeon to harvest skin manually.

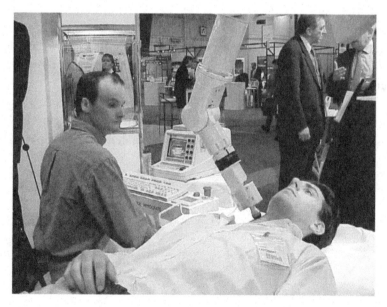

Figure 15.1: Hippocrate: a large cardiovascular disease prevention robot (image adapted from Pierrot et al., 1999)

British-designed robots crawled into the radioactive ruins of the Fukushima Daiichi nuclear power station over the weekend of April 17, 2011, to work in the hazardous debris of the Japanese nuclear disaster, shown in Figure 15.2.

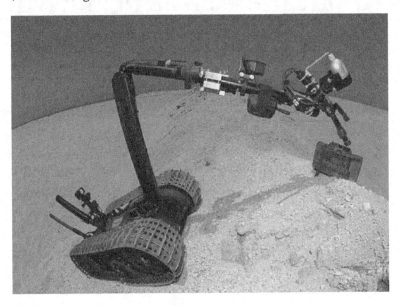

Figure 15.2: Large robots and their trainers in Nuclear Rescue (image is adapted from the new security learning (http://www.newsecuritylearning.com) published by The New Security Foundation (Editor: Dr Harold Elletson) retrieved on March 25, 2013 from http://www.newsecuritylearning.com/index.php/feature/94-robots-and-their-trainers-in-nuclear-rescue)

Dubbed BallP, or Ball Inverted Pendulum, this robot (shown in Figure 15.3) makes use of motors, micro-step controllers, and gyroscopes with incorporated accelerometers to be able to balance upright on the sphere.

Figure 15.3: A BallP micro robot (image adapted from Halme et al., 1996)

The diameter of the ball robot construction is 30 cm, and the weight of the ball robot is only 2.4 kg. This small area contains electronics, which control the motion and the radio connection to the remote (Halme et al., 1996).

The fish micro robot (shown in Figure 15.4) is another miniature device. This robot has actuating and sensing elements installed in it. It can swim at a fast speed in water or in any water medium. This robot is used for a number of medical applications, such as diagnosis and microsurgery with human blood vessels (Fukuda et al., 1995). It also can be used widely in biology and chemistry.

This robot also has many industrial applications, such as inspecting pipes. Many chemical plants, gas supply systems, water supply systems, and heat exchangers use pipelines smaller than a 2 in diameter. These small diameter pipes are not easily accessible for repairs. When a pipe is damaged or leaking, it may contaminate the surrounding soil and ground water; or ground water may penetrate into the pipes. Such pipes are traditionally repaired from above ground. When the damaged pipe portion is detected, the worker will dig a trench to reach and repair the damaged pipe. This traditional approach to repair of pipe systems usually incurs high costs and disruption of traffic and other activity. To overcome these problems, a micro robot can be used for inspecting pipelines that have less than a 2 in diameter. It offers great mobility and provides efficient and practical results (Schreiner, 1994).

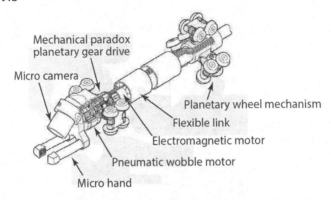

Figure 15.4: Tiny inspection robot (image adapted from Schreiner, 1994)

The tiny inspection robot was developed with a diameter of 22 mm. The length is only 112 mm and the weight is 18 gr. This robot carries a 420000 pixel CCD camera, mounted on the front, which provides high resolution to find minor cracks or damage on the pipe's internal surface. The head of this camera includes an optical power supply, subordinate circuit module, micro optical lens, and filtering glass. The drive signals and power supply for this camera are provided from the cables of the camera controllers. This camera helps in observing the surface of the pipes and recovering the lost parts. It also helps in scaling the pipes. The diameter of this camera is 8 mm, 11 mm in length and the weight is 1.3 gr.

This tiny robot contains a dual hand system, which has 7° of freedom for finding and repairing the obstacles inside the pipe. These hands can carry any object from the internal surface of the pipe, which lies between the range of 0.5 to 4.5 mm in diameter, and are very light in weight, ranging from 1 to 4 gm. The hands are waterproof and dustproof. They are high compliance and easily adaptable to the size of the objects in the pipe.

The hands of the dual hand system as well as the CCD camera can be rotated inside the pipe for circular scanning of the pipe's inner walls, with the help of a pneumatic wobble motor. This motor has a high torque and consists of a stainless steel internal gear, an iron rotor on which an external gear is fabricated, and a wobble ring. It also contains a wobble generator, made of silicon elastic rubber with five chambers (Brunete et al., 2005).

Recently, many flying micro robots have been developed that take their inspiration from a variety of different birds. But, even in comparison to birds, insects can be said to be the most agile flying objects on the earth. Many birds use their wing muscles to fly, and also to change the wing's shape. This results in a change of flight modes and also flapping frequencies. But insects use the elasticity of the skeleton to fly, resulting in a higher flapping frequency as compared to birds. An insect cannot change its flapping frequency because it is fixed by the natural frequency of the insect's external skeleton.

Many investigations have been made of flying insects, to observe their flying mechanisms, wing structure and movements, and drag forces, for application to a tiny insect robot. Figure 15.5 describes the structure of this robot.

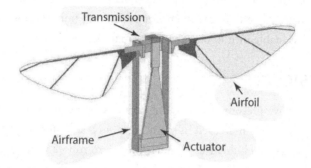

Figure 15.5: Configuration of a tiny insect robot (image adapted from Avadhanula et al., 2003)

The tiny insect-sized robot has mainly four components. The first component is an airframe. It is also called the exoskeleton of the robot. The main functionality of this airframe is to provide solid ground to the actuator during transmission. The second mechanical component is the actuator, also called the flight muscle of this insect robot. The main functionality of this actuator is to provide motion with maximum power density. The third component is the transmission, also called the thorax of the tiny insect robot. The main functionality of this transmission component is to efficiently match the actuator to the load. The last component of this insect robot is the airfoils This is a very important component of this robot; the main functionality of the airfoil component is to remain rigid under large aerodynamic loads, in order to hold its shape (Avadhanula et al., 2003).

The actuator used for this tiny robot is created using a smart-composite, microstructure process. Two important features of this actuator are scalability and compatibility. These actuators, when compared to other actuators, have a high operating stress and frequency capacity. Other major characteristics of this actuator include displacement, bandwidth, and peak force. This actuator acts as the transmission. The weight of the actuator is 50 mg, and the length is 10 mm. The bandwidth of the actuator is more than 1.5 kHz. It creates a bending movement when an electric field is applied, which deflects the transmission (Morrey et al., 2003).

The transmission of this tiny robot amplifies the motion of the actuator, from the input to the output. This process is accomplished by matching the load to the actuator. The wing loading should match with the internal losses of the actuator, to get the efficient electromechanical transduction. The wings are directly controlled by the transmission and the actuators. By adding the additional flexure between the transmission and the wing, the wing can be rotated passively. The airfoils of this robot are designed to match the exact structure the wings. These airfoils are designed very effectively. The wings of this robot provide a high stiffness-to-weight ratio. The airfoils are fixed effectively; the wing of the robot is highly rigid for the expected range of flight forces.

This tiny robot helps in the exploration of hazardous environments, plays a vital role in surveillance, and is helpful for reconnaissance. It has also been used for planetary exploration and for assisting with agriculture. It is useful in the microelectronics and fluid mechanics fields. Finally, this tiny insect robot is very valuable in micromechanics actuations.

15.1 CONCLUSIONS

In this chapter we have presented a sampling of a few micro and macro robots. Recently, the need for micro robots has rapidly increased in many areas such as water supply industries, chemical supply industries, and gas supply industries. Among the micro robots we presented, tiny inspection robots have the most applications. In the future, micro robots are expected to be more widely used for medical applications, such as diagnosis and surgery (Poignet et al., 2003), so that mankind is more likely to benefit directly.

References

Aggarwal, A. (1984). The art gallery theorem: Its variations, applications, and algorithmic aspects. Ph.D. thesis, Johns Hopkins University. 94

Albus, J. S. and Meystel, A. (1996). A Reference Model Architecture for Design and Implementation of Intelligent Control in Large and Complex Systems. International Journal of Intelligent Control and Systems, 1(1). (pp. 15–30). Westing Publishing Co. DOI: 10.1142/S0218796596000039. 32

Antich, J. and Ortiz, A. (2009). Bug2+: Details and formal proofs. Technical Report A-1-2009, University of the Balearic Islands. The paper can be downloaded from http://dmi.uib.es/ ~jantich/Bug2+.pdf. 90

Applegate, D.L., Bixby, R.E., Chvátal, V., and Cook, W.J. (2007). The Traveling Salesman Problem: A Computational Study. Princeton University Press. 94

Arkin, R. (1998). Behavior Based Robotics. MIT Press. 1, 21

Avadhanula, S., Wood, R.J., Steltz, E., Yan, J., and Fearing, R.S. (2003). Lift force improvements for the micromechanical flying insect. Proceedings of the International Conference on Intelligent Robots and Systems (IROS). Piscataway, NJ: IEEE Press. (pp. 1350–1356). 131

Ayala, F. (2010). Am I a Monkey? Johns Hopkins University Press. 9, 10

Bajcsy, R. (1995). From active perception to active cooperation - fundamental processes of intelligent behavior. University of Pennsylvania, MS-CIS-95-07. 45

Balch, T. (1998). Behavioral Diversity in Learning Robot Teams. Ph.D. Thesis, College of Computing, Georgia Institute of Technology. 103

Balch, T. and Arkin, R. (1998). Behavior-Based Formation Control for Multi-Robot Teams. IEEE Transactions on Robotics and Automation, XX(Y). DOI: 10.1109/70.736776. 103

Balch, T. and Parker, L. (2002). Robot Teams: From Diversity to Polymorphism. AK Peters, Ltd. 103

Balcombe, J. (2010). Second Nature: the Inner Lives of Animals. Palgrave Pub. 22

Bar-Cohen, Y. and Breazeal, C. (2003). Biologically Inspired Intelligent Robotics. Proceedings of the SPIE Smart Structures Conference. 104

Begall, S., Červený, J., Neef, J., Vojtěch, O., and Burda, H. (2008). Magnetic Alignment in Grazing and Resting Cattle and Deer. Proceedings of National. Academy of Science, 105(36) pp. 13451–13455, NAS press. DOI: 10.1073/pnas.0803650105. 22

Bekey, G. (2005). Autonomous Robots: From Biological Inspiration to Implementation and Control. MIT Press. 1

Bonabeau, E. (1999). Swarm Intelligence: From Natural to Artificial Systems. Oxford University Press. 21

Bonasso, R. P. and Myers, K. L. (1998). The Home-Vacuum Event. AI Magazine 19(3) 29–32, MIT press. 32

Bonasso, R. P., Firby, R. J., Gat, E., Kortenkamp, D., Miller, D.P., and Slack, M. G. (1997). Experiences with an Architecture for Intelligent Reactive Agent. Journal of Experimental and Theoretical Artificial Intelligence JETAI, 9. 32

Braitenberg, V. (1984). Vehicles - Experiments in Synthetic Psychology. MIT press. 22

Breazeal, C. (1998). Learning by Scaffolding. Ph.D. Thesis Proposal, M.I.T. Department of Electrical Engineering and Computer Science. 41

Brooks, R. (1985). AI Memo 864: A Robust Layered Control System For a Mobile Robot. MIT. 21, 28, 31, 32

Brooks, R. (1986). A Robust Layered Control System for a Mobile Robot. IEEE Journal of Robotics and Automation, IEEE Press. DOI: 10.1109/JRA.1986.1087032. 31

Brown, E., Rodenberg, N., Amend, J., Mozeika, A., Steltz, E., Zakin, M., Lipson, H., and Jaeger, H. (2010). Universal robotic gripper based on the jamming of granular material. Proceedings of the National Academy of Sciences (cover), 107(44), pp. 18809–18814. DOI: 10.1073/pnas.1003250107. 59

Brunete, A., Hernando, M., & Gambao, E. (2005). Modular multiconfigurable architecture for low diameter pipe inspection microrobots. Proceedings of the International Conference on Robotics and Automation (ICRA). Piscataway, NJ: IEEE Press, pp. 490–495. 130

Carbonell, J.R. (1970). AI in CAI: An artificial intelligence approach to computer-assisted instruction. IEEE Transactions on Man-Machine Systems, 11(4), pp. 190–202, IEEE. DOI: 10.1109/TMMS.1970.299942. 37

Carlsson, S., Jonsson, H., and Nilsson, B.J. (1993). Finding the shortest watchman route in a simple polygon. Springer. 94

Chakraborty, B. (2008). Feature Subset Selection by Particle Swarm Optimization with Fuzzy Fitness Function. 3rd International Conference on Intelligent System and Knowledge Engineering, pp. 1038–1042. 115

Chalmers, D. (1996). The Conscious Mind: In Search of a Fundamental Theory. Oxford University Press. 37

Choset, H., Lynch, K.M., Hutchinson, S., Kantor, G.A., Burgard, W., Kavraki, L.E., and Thrun, S. (2005). Principles of Robot Motion: Theory, Algorithms, and Implementations. MIT press. 1, 2, 77, 85, 87, 88, 89, 91, 93

Chung, T.H., Hollinger, G., and Isler, V. (2011). Search and Pursuit-Evasion. Mobile Robotics, pp. 299–316, Springer. 120

Cipolla, R. and Hollinghurst, N.J. (1996). Human-robot interface by pointing with uncalibrated stereo vision. Image and Vision Computing, 14(3), pp. 171–178, Elsevier Pub. DOI: 10.1016/0262-8856(96)84056-X. 110

Conitzer, V. (2011). Solving Stackelberg Games with Uncertain Observability. Proc. of 10th International Conference on Autonomous Agents and Multiagent Systems (AAMAS), ACM. 121

Connell, J. (1989). A Colony Architecture for an Artificial Creature. MIT Ph.D. Thesis in Electrical Engineering and Computer Science, MIT. 32

Cormen, T.H., Leiserson, C.E., Rivest, R.L., and Stein, C. (2001). Introduction to Algorithms. MIT press. 71

Daniels, J., Brower, J., and Baumgartner, J. and F. (1998). High resolution GPR at Brookhaven National Laboratory to delineate complex subsurface targets. Journal of Environmental and Engineering Geophysics, 3(1), pp. 1–5. DOI: 10.4133/JEEG3.1.1. 50

Dankers, A. and Zelinsky, A. (2004). CeDAR: A Real-World Vision System. *Machine Vision and Applications manuscript.* 16(1), pp. 47-58. DOI: 10.1007/s00138-004-0156-3. 52

Dasarathy, B.V. (1998). Sensor Fusion: Architectures, Algorithms, and Applications II. Proceedings of SPIE conference, IEEE. 47

Dautenhahn, K., and Werry, I. (2004). Towards interactive robots in autism therapy: Background, motivation and challenges. Pragmatics & Cognition, 12(1), pp. 1–35. DOI: 10.1075/pc.12.1.03dau. 38

Denavit, J. and Hartenberg, M. (1955). A Kinematic Notation for Lower-Pair Mechanisms Based on Matrices. Journal of Applied Mechanics, June, pp. 215–221. 55

Dennett, D. (1989). The intentional Stance. MIT press. 37

de Silva, C.W. (1995). Intelligent Control: Fuzzy Logic Applications. CRC press. 2

132 REFERENCES

Dias, M.B. and Ollero, A., Guest editors (2007).. Teamwork in Field Robotics. Journal of Field Robotics, 24(11, 12) Wiley. DOI: 10.1002/rob.20230. 111

Di Lorenzo, G., Pinelli, F., Pereira, F.C., Biderman, A., Ratti, C., and Lee, C. (2009). An Affective Intelligent Driving Agent: Driver's Trajectory and Activities Prediction. Vehicular Technology Conference, IEEE. 40

Dorigo, M. (2004). Ant Colony Optimization. MIT Press. 105

Dudek, G. and Jenkins, M. (2010). Computational Principles of Mobile Robotics. Cambridge University Press. 1, 74

Dunias, P. (1996). Autonomous robots using artificial potential fields. Technische Universiteit Eindhoven. 66

Eberhart, R. C., Shi, Y., and Kennedy, J. (2001). Swarm Intelligence. Morgan Kaufmann pub. 105

Everett, H. R. (1995). Sensors for Mobile Robots. A K Peters/CRC Press. 45

Fautin, D.G. (1991). The Anemonefish Symbiosis: What is Known and What is not, Journal of Symbiosis, 10, pp. 23–46, Springer Press. 10

Ferranti, E., Trigoni, N. and Levene, M. (2007). Brick & Mortar: An On–Line Multi–Agent Exploration Algorithm. ICRA07, IEEE International Conference on Robotics and Automation. IEEE. DOI: 10.1109/ROBOT.2007.363078. 97, 98

Ferranti, E. and Trigoni, N. (2011). Practical Issues in Deploying Mobile Agents to Explore a Sensor–Instrumented Environment. The Computer Journal, 54(3), pp. 309–320. DOI 10.1093/comjnl/bxq013. 98

Floreano, D. and Mattiussi, C. (2009). Bio-Inspired Artificial Intelligence: Theories, Methods, and Technologies. MIT Press. 117

Fukuda, T., A. Kawamoto, and F. Arai, (1995). Steering mechanism of underwater micro mobile robot. Proceedings of the International Conference on Robotics and Automation. Piscataway, NJ: IEEE Press, pp. 363–368. 129

Franklin, S. P. (1995). Artificial Life, in Artificial Minds. MIT Press. 21

Gavrilova, M. (2008). Generalized Voronoi Diagram: A Geometry-Based Approach to Computational Intelligence. Springer. 75

Gazi, V., Passino, K.M. (2003). Stability Analysis of Swarms. IEEE Trans. on Automatic Control, 48(4), pp. 692–697, IEEE Press. 27

Gee, H. (2008). Evolutionary Biology: The amphioxus unleashed. Nature 453, pp. 999–1000. DOI: 10.1038/453999a. 10

González, E., Álvarez, O., Díaz, Y., Parra, C., and Bustacara, C. (2005). BSA: A Complete Coverage Algorithm. Proceedings of the 2005 IEEE International Conference on Robotics and Automation, pp. 2040, 2044, IEEE. DOI: 10.1109/ROBOT.2005.1570413. 93

Goodrich, M.A. (2000). Potential Fields Tutorial. http://www.ee.byu.edu/ugrad/srprojects/robot-soccer/papers/goodrich_potential_fields.pdf. 64, 65

Gould, J.L. (1982). Ethology: The Mechanisms and Evolution of Behavior. W. W. Norton & Company. 21

Govindan, R. and Tangmunarunkit, H. (2000). Heuristics for Internet map discovery. Infocom proceedings, 3, PP. 1371–1380, IEEE. 67

Grinde, B. (2002). Darwinian Happiness: Evolution as a Guide for Living and Understanding Human Behavior. The Darwin Press, Inc. 22

Haller, S., McRoy, S., and Kobsa, A. (1999). Computational Models of Mixed-Initiative Interaction. Kluwer Pub. 37

Halme, A., Schönberg, T., and Wang, Y. (1996). Motion Control of a Spherical Mobile Robot Mie University, Japan, IEEE. 129

Harper, P.G. and Weaire, D.L. (1985). Introduction to Physical Mathematics. Cambridge University Press. 63

Hashimoto, T., Hitramatsu, S., Tsuji, T. and Kobayashi, H. (2006). Development of the Face Robot SAYA for Rich Facial Expressions. SICE-ICASE International Joint Conference, IEEE. 41

Hayes, S.T., Hooten, E. R., and Adams, J.A. (2010). Multi-touch interaction for tasking robots. Proceedings of the 5th ACM/IEEE International Conference on Human-Robot Interaction, pp. 97–98, IEEE. 109

Hershberger, W. H. (1989). Volitional Action. North Holland. 8

Hexmoor, H., Kortenkamp, D., Horswill, I. (1997). Software architectures for Hardware Agents. J. Experimental and Theorretical Artificial Intelligence, 9(2-3), Taylor and Francis Pub. 29

Hexmoor, H. and Shapiro, S. (1997). Integrating Skill and Knowledge in Expert Agents. Expertise in Context, Feltovich, Ford, and Hoffman (Eds), 383, 404, AAAI/MIT Press. 32, 33

Hexmoor, H. and Bandera, C. (1998). Architectural Issues for Integration of Sensing and Acting Modalities. ISIC/CIRA/ISAS 1998, IEEE press. 52

Hexmoor, H. and Desiano, S. (1999). Autonomy Control Software. The Knowledge Engineering Review (1999), 14 : pp 377–382, Cambridge University Press. 29

Hexmoor, H. and Yang, J. (2000). Pointing: A Component of a Multimodal Robotic Interface. Proceedings of the Workshop in Interactive and Entertainment Robots (Wire-2000), p. 103–107, CMU. 110, 111

Hexmoor, H., Castelfranchi, C., Falcone, R. (2003). Agent Autonomy. Springer. 8

Hexmoor, H., Rahimi, S., and Chandran, R. (2008). Delegations guided by trust and autonomy. Web Intelligence and Agent Systems 6(2): 137–155, IOS Press. 103

Ho, C.C., MacDorman, K.F., and Dwi Pramono, Z. A. D. (2008). Human emotion and the uncanny valley: a GLM, MDS, and Isomap analysis of robot video ratings. Proceedings of the 3rd ACM/IEEE international conference on Human robot interaction, ACM Press. DOI: 10.1145/1349822.1349845. 42

Homaifar, A., Battle, D., and Tunstel, E. (1999). Soft computing-based Design and Control for Mobile Robot Path Tracking. IEEE International Symposium on Computational Intelligence in Robotics and Automation. CIRA'99, pp. 35–40. 117

Horiuchi, Y. and Noborio, H. (2001). Evaluation of Path Length Made in Sensor-Based Path-Planning with the Alternative Following. Proc. of the 2001 IEEE International Conference on Robotics and Automation, pp.1728–1735, IEEE Press. 90

Horvitz, E.(1999). Uncertainty, Action, and Interaction: In Pursuit of Mixed-Initiative Computing. Intelligent Systems, September 1999, IEEE Computer Society. 37

Huang, J., Di, P., Fukuda, T., and Matsuno, T. (2008). Motion Control of Omni-directional Type Cane Robot-based on Human Intention. Proc. International Conference on Intelligent Robots and Systems (IROS), pp. 273–278, IEEE. 121

Humphrey, C.M. and Adams, J.A. (2011). Analysis of complex, team-based systems: Augmentations to goal-directed task analysis and cognitive work analysis. Theoretical Issues in Ergonomics, 12(2), pp.149–175. DOI: 10.1080/14639221003602473. 109

Jacobsen, S., Iversen, E., Knutti, D., Johnson, R., and Biggers, K, (2003). Design of the Utah/M.I.T. Dextrous Hand. Proc. of Robotics and Automation. 1986 IEEE. pp. 1520–1532. 58

Kaelbling, L.P. (1993). Learning in Embedded Systems. MIT Press. 26

Kamon, I., Rivlin, E., and Rimon, E. (1998). TangentBug: A range-sensor based navigation algorithm. Journal of Robotics Research, 17(,) pp. 934–953. DOI: 10.1177/027836499801700903. 90

Kernbach, S., Meister, E., Schlachter, F., Jebens, K., Szymanski, M., Liedke, J., Laneri, D., Winkler, L., Schmickl, T., Thenius, R., Corradi, P., and Ricotti, L. (2008). Symbiotic Robot Organisms: Replicator and Symbrion projects, PerMIS. ACM Press. 28

Kleinberg, J. (2005). Algorithm Design. Addison-Wesley. 67

Kim, D., Song, W., Han, J., and Bien, Z.Z. (2003). Soft Computing based Intention Reading Techniques as a Means of Human-robot Interaction for Human Centered System. Soft Computing - A Fusion of Foundations, Methodologies and Applications, 7, pp. 160–166. 118

Kirby, B., Goldstein, S., Mowry, T., Aksak, B., and Hoburg, J. (2007). A Modular Robotic System Using Magnetic Force Effectors. Proceedings of the IEEE International Conference on Intelligent Robots and Systems (IROS '07). 106

Kling, R., Rosenbaum, H., and Sawyer, S. (2005). Understanding and Communicating Social Informatics: A Framework for Studying and Teaching the Human Contexts of Information and Communications Technologies. Medford, New Jersey: Information Today, Inc. 106

Knight, H. (2011). Eight Lessons learned about Non-verbal Interactions through Robot Theater. International Conference on Social Robotics. DOI: 10.1007/978-3-642-25504-5_5. 42

Lakov, D. (2002). Fuzzy Agents versus Intelligent Swarms. First International IEEE Symposium Intelligent Systems, IEEE. DOI: 10.1109/IS.2002.1044239. 115

Lammens, J., Hexmoor, H., and Shapiro, S. (1993). Of Elephants and Men. The Biology and Technology of Intelligent Autonomous Agents, Luc STeels (Ed.), pp. 312–344, Springer. 32

Latombe, J.C. (1991). Robot Motion Planning. Kluwer Academic Publishers. 63, 67, 76, 77, 85

LaValle, S. M. and Hutchinson, S.A. (1993). Game Theory as a Unifying Structure for a Variety of Robot Tasks. Proc. International symposium on intelligent control, pp. 429–434, IEEE. DOI: 10.1109/ISIC.1993.397675. 119

Lewis, D. (1970). Polynesian and Micronesian Navigation Techniques. Journal of Navigation, 23, pp. 432–447, The Royal Institute of Navigation. DOI: 10.1017/S0373463300020683. 86

Lin, P., Abney, K., Bekey, G. A. (Editors) (2011). Robot Ethics: The Ethical and Social Implications of Robotics. MIT Press. 1, 7

Lipson, H. (2008). Evolutionary Synthesis of Kinematic Mechanisms, Artificial Intelligence in Design and Manufacturing, 22, pp. 195–205. 20

Lockwood, J. (2009). Sex-Leged Soldiers. Oxford University Press. 16

Long, J. (2012). Darwin's Devices. Basic Books Pub. 9

Lorini, E., Piunti, M., Castelfranchi, C., Falcone, R., and Miceli, M. (2008).Anticipation and Emotions for Goal Directed Agents, In The Challenge of Anticipation, Lecture Notes in Computer Science, 2008, 5225/2008, pp. 135–160, Springer. 38

Lozano-Pérez, T. and Wesley, M. A. (1979). An algorithm for planning collision-free paths among polyhedral obstacles. Communications of the ACM 22(10): pp. 560–570, ACM. DOI: 10.1145/359156.359164. 68

Lozano-Perez, T. (1983). Spatial Planning: A Configuration Space Approach. IEEE Transactions on Computers, C-32(2), pp. 108–120. DOI: 10.1109/TC.1983.1676196. 85

Lumelsky, V. and Stepanov, A. (1987). Path planning strategies for a point Mobile Automaton Moving amidst Unknown Obstacles of Arbitrary Shape. Algorithmica, 2, pp. 403–430, Springer. DOI: 10.1007/BF01840369. 86

McLaughlan, B., Hexmoor, H. 2009. Influencing Massive Multi-agent Systems via Viral Trait Spreading. Third IEEE International Conference on Self-Adaptive and Self-Organizing Systems. 27, 104

Meng, Y. (2008). Multi-Robot Searching using Game Theory Based Approach. International Journal of Advanced Robotics, 5(4), pp. 341–350. DOI: 10.5772/6232. 119

Miller, D.P., Tan, L., and Swindell, S. (2003). Simplified Navigation and Traverse Planning for a Long-Range Planetary Rover. Proc. of International conference on Robotics and Automation, pp. 2436–2441, IEEE Press. 82

Minsky, M. (1998). Society of Mind. Simon and Schuster Pub. 22

Morrey, J. M., Lambrecht, B. G. A., Horchler, A. D., Ritzmann, R. E., and Quinn, R. D. (2003). Highly Mobile and Robust Small Quadruped. Proceedings of International Intelligent Robots and Systems (IROS), pp. 82–87, IEEE Press. 131

Munkres, J.R. (2000). Topology. Prentice Hall. 67

Ng, J (2010). An Analysis of Mobile Robot Navigation Algorithms in the Unknown Environments. Ph.D. thesis, University of Western Australia. 88, 89

Nocks, L. (2007). The Robot: The Life Story of a Technology. Greenwood pub. 10

Ortony, A., Clore, G.L., and Collins, A. (1990). The Cognitive Structure of Emotions. Cambridge University Press. 41

Paruchuri, P., Pearce, J., and Kraus, S. (2008). Playing Games for Security: An Efficient Exact Algorithm for Solving Bayesian Stackelberg Games. Proc AAMAS. 121

Paul, R. (1981). Robot Manipulators: Mathematics, Programming, and Control. MIT Press. 53

Picard, R. (1997). Affective Computing. MIT Press. 37

Pierrot, F., Dombrea, E., Dégoulangeb, E., Dégoulange, E., Urbain, L., Caron, P., Gariépy, J., and Mégnien, J.L. (1999). Hippocrate: a Safe Robot Arm for Medical Applications with Force

Feedback. Journal of Medical Image Analysis, 3(3), pp. 285–300, Elsevier Pub. DOI: 10.1016/S1361-8415(99)80025-5. 127

Pitta, J., Tambe, M., Kiekintveld, C., Cullen, S., and Steigerwald, E. (2011). Game Theoretic Security Allocation on a National Scale. AAMAS. 122

Plomin, R., Owen, M.J., and McGuffin, P. (1994). The genetic basis of complex human behaviors. Journal Science, 264(5166), pp. 1733–1739, American Association for Advancement of Science. 21

Poignet, P., Dombre, E., Merigeaux, O., and Pierrot, F. (2003). Design and Control Issues for Intrinsically Safe Medical Robots. Journal of Industrial Robot, 30(1), pp.83–88, Emerald Pub. DOI: 10.1108/01439910310457751.132

Power, G.H. (1992). How might Emotions Affect Learning?, In The handbook of emotion and memory, pp. 3–31, Lawrence Erlbaum Associates. 38

Raibert, M. (1986). Legged Robots, Communications of the ACM, 29(6), ACM. 16

Raiffa, H. (1968). Decision Analysis: Introductory Lectures on Choices under Uncertainty. Addison Wesley. 119

Rao, N.S.V. (1995). Robot navigation in unknown generalized polygonal terrains using vision sensors. IEEE Transactions on Systems, Man and Cybernetics, 26(6), pp. 947–962, IEEE. DOI: 10.1109/21.384257. 82

Rawlinson, D. and Jarvis, R. (2007). Topologically-directed navigation. Robotica, pp. 1 of 15, Cambridge University Press. 72

Resnick, M. (1997). Turtles, Termites, and Traffic Jams: Explorations in massively parallel microworlds. MIT Press. 22

Saffiotti, A. (1997). The Uses of Fuzzy Logic in Autonomous Robot Navigation. Soft Computing, 1, pp. 180–197. DOI: 10.1007/s005000050020. 115, 116

Salumäe, T., Rañó, I., Akanyeti, O., and Kruusmaa, M. (2012). Against the flow: A Braitenberg controller for a fish robot. Proc. International Conference on Robotics and Automation (ICRA) pp. 4210–4215, IEEE. 24

Schreiner, S. I. (1994). Internal Pipe Inspection Devices for Use in Radiation Survey Applications. Transactions of the American Nuclear Society, 71(1), pp. 514–515, MIT Press. 129, 130

Seung, S. Connectome. Houghton Mifflin Harcourt Pub. 20

Shannon, C.E. and Weaver, W. (1971). The Mathematical Theory of Communication. University of Illinois Press. 103

Shen, D., Genshe, C., Cruz, J.B., and Blasch, E. (2008). A Game Theoretic Data Fusion Aided Path Planning Approach for Cooperative UAV ISR. Aerospace Conference, IEEE Press. 98

Shubin, N. (2008). Your Inner Fish: A Journey into the 3.5-Billion-Year History of the Human Body. Pantheon Pub. 9

Siegwart, R., Nourbakhsh, I. (2011). Introduction to Autonomous Mobile Robots. MIT Press. 15, 75

Skinner, B.F. (1977). Why I am Not a Cognitive Psychologist. Behaviorism, 5(2), pp. 1–10, Cambridge University Press. 37

Slack, M.G. (1993). Navigation Templates: Mediating Qualitative Guidance and Quantitative Control in Mobile Robots. IEEE Trans. On System, Man, and Cybernetics, 23(2), IEEE Press. 82

Spong, M. and Vidyasagar, M. (1989). Robot Dynamics and Control. Wiley Press. 55

Steunebrink, B., Dastani, M., and Meyer, J.J. (2007). Emotions as Heuristics for Rational Agents. Technical Report UU-CS-2007-006, Utrecht University. Available online at www.cs.uu.nl. 109

Symington, A., Waharte, S., Julier, S., and Trigoni, N. (2010). Probabilistic Target Detection by Camera-Equipped UAVs. IEEE Conference in Robotics and automations, IEEE Press. 98

Takahashi, Y. andTakagi, H. (200)7. Simple Biped Walking Robot for University Education Considering Fabrication Experiences, Autonomous Robots and Agents. Studies in Computational Intelligence, 2007, 76/2007, pp. 181–18, Springer. 42

Taylor, T. (2010). The Artificial Ape: How Technology Changed the Course of Human Evolution. Palgrave Macmillan. 9, 22

Thrun, S., Burgard, W., and Fox, D. (2005). Probabilistic Robotics. MIT Press. 2

Tolman, E.C. and Honzik, C.H. (1930). "Insight" in rats. University of California Publications in Psychology, 4, 215–232. 22

Tolman, E.C., Ritchie, B.F. and Kalish, D. (1946). Studies in spatial learning: I. Orientation and the short-cut. Journal of Experimental Psychology. 36, 13–24. DOI: 10.1037/h0053944. 22

Trappl, R., Payr, S. (2009). Agent Culture: Human-Agent Interaction in a Multicultural World. LEA Press. 8

Tunstel, E., Akbarzadeh-T, M., Kumbla, K., and Jamshidi, M. (1996). Soft Computing Paradigms for Learning Fuzzy Controllers with Applications to Robotics. Proceedings of North American Fuzzy Information Processing, pp. 355–359. DOI: 10.1109/NAFIPS.1996.534759. 117, 118

Turocy, T.L. and von Stengel, B. (2002) Game Theory. Academic Press. 119

Vallverdú, J. and Casacuberta, D. (2009). Handbook of Research on Synthetic Emotions and Sociable Robotics: New Applications in Affective Computing and Artificial Intelligence. IGI Pub. 42

Waltz, E.L., Llinas, J., and White, F.E. (1990). Multisensor Data Fusion. Artech House Publishers. 47

Wehrle, T. and Scherer, K.R. (2001). Towards Computational Modeling of Appraisal Theories. In Appraisal processes in emotions: Theory, methods, research, K. R. Scherer, A. Schorr, & T. Johnstone (eds), pp. 350–365, Oxford University Press. 40

Wendt, C. and Berg, G. (2009). Nonverbal Humor as a New Dimension of HRI. In Proc. 18th IEEE International Symposium on Robot and Human Interactive Communication, pp. 183–188, IEEE. 42

Wenninger, M. J. (1974). Polyhedron Models. Cambridge University Press. 68

Wooldridge, M. (2009). An Introduction to MultiAgent Systems. John Wiley & Sons. 29, 104

Woolfson, M. M. (2008). Time, Space, Stars and Man: The Story of the Big Bang. Imperial College Press, UK. 9

Yanco, H.A. (2000). Shared User-Computer Control of a Robotic Wheelchair System. Ph.D. Thesis, Department of Electrical Engineering and Computer Science, Massachusetts Institute of Technology. 107

Yanco, H. and Drury, J. (2004). Classifying Human-Robot Interaction, Systems: An Updated Taxonomy, Man and Cybernetics, IEEE International Conference. 107

Young, S.E., Miller, R., McDonald, S.S. (2007). Keys to Innovative Transport Development. Presented at the 87 Annual Meeting of the Transportation Research Board, Washington, D.C. 15

Zadeh, L. (1965). The Berkeley Initiative on Soft Computing. http://www.eecs.berkeley.edu/~zadeh 113

Author Biography

Henry Hexmoor received his M.S. degree from Georgia Tech, Atlanta, and his Ph.D. degree in computer science from the State University of New York, Buffalo, in 1996. He is an IEEE senior member.

Henry has taught at the University of North Dakota before a stint teaching at the University of Arkansas. Currently, he is an associate professor with the Computer Science Department, Southern Illinois University, Carbondale, IL. He has published widely on artificial intelligence and multiagent systems. His research interests include multiagent systems, artificial intelligence, cognitive science, mobile robotics, and predictive models for transportation systems.